Introduction to the Wildflowers of Northern Ireland

James Napier

Author biography

Dr James Napier has an Honours degree and a Master of Science in Biological Science. He completed his Doctorate at the University of Ulster on the evolutionary ecology of Wild Garlic.

James is the author of a wide range of textbooks for GCSE and A-level students in Science and Biology, and he has also written several Popular Science books covering the topics of Genetics and Evolution.

He is also the author of books on the topics of attention deficit disorder and prostate cancer, all the proceeds of which were donated to mental health and cancer charities.

First published in 2022 by James Napier

Text and photographs © James Napier, 2022
All rights reserved

James Napier has asserted his right under
the Copyright, Designs and Patents Act 1988 to be
identified as the author of this work.

Printed by W&G Baird Ltd, Antrim

ISBN 978-1-3999-2897-7

Copies of this book can be ordered from Colourpoint Books, Newtownards
or directly from the author at napierflowers@gmail.com

Cover photograph: Early-purple Orchids in grassland on the Belfast hills.

Contents

Acknowledgements

Many thanks to Bob Davidson for reviewing the manuscript in detail and providing advice and suggestions throughout. Bob has been involved with biodiversity and conservation throughout his life; his expertise as a field botanist of note has proved invaluable in the completion of this book.

Thanks to Dr Reginald Haslett for his expertise and time in reviewing and editing the photographs. Reggie is a well-known and accomplished photographer and his advice and support in this area is much appreciated.

My former colleague Ronnie Irvine has my thanks for his role in introducing me to some excellent plant habitats in Northern Ireland, and particularly, for stimulating my interest in our native Orchids. An excellent field botanist and photographer himself, his advice in these areas is always welcome.

Introduction

Northern Ireland is home to some beautiful formal gardens with glorious floral displays which attract visitors in their thousands each year. It is also home to equally good, if not better, displays of native wildflowers in their natural settings.

Some wildflowers are so common that they can occur virtually anywhere; think **Dandelion** or **Daisy** – recognisable to virtually all to the extent that they seldom get a second look. Children can identify Dandelions and Daisies with ease; Daisy chains and Dandelion 'clocks' are part of all our experiences.

The ubiquitous Daisy

Other species (types of plants) are less well known but common in their typical habitat. If you know where to look and, crucially, know when it is the right time of the year to seek them out, you have a good chance of finding them in their natural setting. A good example of a species in this category is Thrift but there are many others.

Thrift (also known as Sea Pink) is a species common on rocky shores and sea cliffs. The presence of many plants and the many beautiful powder-pink flowers on each plant give sections of the Northern Ireland (NI) coast a pink hue from April to June. Thrift is probably the flowering plant most associated with the coast for many people.

Thrift in its typical habitat –
the spray zones of rocky shores or on sea cliffs

The **Green-winged Orchid** is a species in another category altogether. Unlike Thrift, this beautiful plant is extremely rare in Northern Ireland, occurring in only one small part of the County Down coast. At this locality, it only occurs in an area covering scores rather than hundreds of square metres. The next photograph showing this Orchid was taken early morning in late May, by which time the Sun had risen just enough to bathe these exquisite plants in sunshine. Not surprisingly, the delicate flowers of this species only last for a few weeks, and less if the Spring weather is severe. More later about this rare and stunning plant.

The Green-winged Orchid in its only Northern Ireland site is in flower during May and early June

The plant kingdom is a major grouping in the living world and within this kingdom flowering plants form one of several large groupings. Individual species covered in this book will be referred to by their common name (or names), rather than using scientific names.

The next chapter has a short introduction on flower structure and the role of flowers in plants. Most of the rest of the book is on flowering plants in their natural settings and describes a sample of the plants you would expect to come across in our major habitats including coasts, woodland, wetland, grassland, heath and bog. This is followed by a chapter on a year in the life of a hedgerow. Poisonous species and aliens (in the plant world) are referred to as appropriate – some surprises there! In addition, there is a glossary of botanical terms at the end of the book.

But what really is the function of this book? Being an 'introduction', it should be meaningful to anyone with an interest in wildflowers and even those with just a general enjoyment of the outdoor environment. Those who do not classify themselves as expert botanists but would like to know more about our plant species and where to find them, are probably the main target audience.

Hopefully, you will find this tour through the wildflowers of Northern Ireland interesting, informative and enjoyable.

Flowers and Flowering Plants

In simple terms, flowers are the reproductive organs of flowering plants. Mosses and ferns are plants, but they do not produce flowers. Nor do conifers – flowering is only one of several types of reproductive strategy in the plant kingdom. Consequently, these plants are not included in the large plant group known as flowering plants. Many of our common trees such as Oak and Ash are technically flowering plants, but for the purposes of this book the emphasis will be on what most of us understand as flowering plants – the herbaceous (non-woody) species.

The overriding function of a flower is reproduction via seed production, a normal outcome if flowers do their job. When conditions are suitable, seeds can germinate to produce new plants and the cycle continues. Additionally, seeds can be dispersed to new areas allowing a species to spread and colonise new places. It is convenient to think of a seed as the initial stage in the life of a flowering plant. But you need flowers to get seeds, so how does this happen?

The structure of a flower

Flowers come in all shapes, colours and sizes – to understand the parts of a flower it is best to consider the structure of a simple flower such as the flower of **Lesser Celandine**, a species closely related to the Buttercups and common in woodland and hedgerows in early Spring.

At a basic level, flowers can be considered as a series of structures arranged in concentric rings or variations of this theme. The outermost ring usually contains the **sepals**. Sepals are often green and typically form a protective layer around the **petals** (the next 'ring' moving towards the flower centre), which are often brightly coloured. The collective name for the sepals is the **calyx** and the petals collectively are referred to as the **corolla**.

In Lesser Celandine, there are three green sepals and usually between eight and twelve yellow petals.

Sometimes petals and sepals are visually indistinguishable. In this situation, the petals and sepals are together referred to as **tepals**, with the collective term being **perianth**.

Within the ring of petals are the parts which produce the sex cells – the sperm and eggs of the plant world. The stamens are arranged in a ring around the central carpels. The **stamens** are the male parts, with each stamen consisting of a long stalk (**filament**) and the **anther** (usually at the top). It is the anther which produces the **pollen**, the structure containing the male sex cell or gamete. Centrally is situated one or more carpels, the part which houses the female sex cells. Each **carpel** consists of an upper **stigma** (the structure upon which the pollen lands during pollination) and a **style** which runs from the stigma to the (usually) swollen bottom section of the carpel called the **ovary**. It is the ovary which contains one or more **ovules** or 'eggs'.

The diagram below represents a generalised flower, showing how the different components relate to each other.

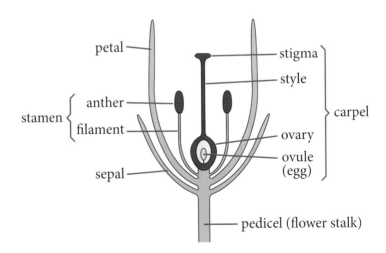

A generalised flower

The photograph below of a Lesser Celandine flower shows some of the structures referred to earlier.

Flower of Lesser Celandine showing petals, stamens and carpels.
The sepals cannot be seen as they are behind the petals.

Looking closely at the stamens in the photograph, it is possible to distinguish between the anthers which produce the pollen and the filaments or stalks which raise the anthers above the centrally positioned carpels in this species.

Other flower structures will be considered later in this chapter, but first it is useful to consider how reproduction takes place in flowers.

Pollination and fertilisation in flowers

Think of reproduction in flowering plants as a two-part process. **Pollination** is the transfer of pollen grains from anther to stigma. Once pollination is completed, the nucleus of the pollen grain travels down through the stigma and style of the carpel and enters the ovary. Once this is achieved, the pollen nucleus and the nucleus of the ovule combine (fuse) in the process of **fertilisation** to form the embryo – analogous to the fusion of the sperm nucleus and the egg nucleus in human fertilisation. The embryo then develops into a seed. It is more complex than this of course – pollination doesn't guarantee fertilisation, but if there is no pollination there will be no fertilisation!

Self and Cross-pollination

Self-pollination is pollination within the same flower (or between flowers of the same plant).

Cross-pollination involves pollen being transferred from the stamen(s) in the flower of one plant to the stigma in a flower of a *different* plant of the same species.

As cross-pollination involves greater genetic mixing (outbreeding), it is favoured from an evolutionary perspective and many plant species have strategies to encourage cross-pollination. Some species use the most certain method of avoiding self-pollination by having plants which are either male or female. The flowers of male plants have stamens but no carpels, and the flowers of female plants have carpels but no stamens. Examples of species with separately sexed plants include Willows, the Common Nettle and Red Campion.

However, most species do have flowers with both male and female parts and are hermaphroditic. In these species, there are many effective strategies for preventing or reducing the likelihood of self-pollination.

One such strategy involves the stamens and carpels maturing

Male catkins in Willow.
The flowers shown have stamens and very little else. The smaller
catkins on either side of the larger central ones have not yet opened.
A catkin is a cylindrical cluster of small flowers without petals or
have petals which are very indistinct. Examples of species with
flowers in catkins include the Willows, Alder and Hazel.

at different times within the same flower. This is a common
feature – much more so than having separately sexed plants –
and many of the species described in this book have developed
this strategy.

Rosebay Willowherb, a coloniser of cleared woodlands, waste ground and many other habitats, has flowers in which the stamens mature before the carpels. The photograph below shows that flowers in this species have four dark purple pointed sepals, within which are four petals notched at their flatter tips, which are of a lighter mauve-purple colour. The stamens are arranged around a central carpel.

The eight stamens of Rosebay Willowherb have large terminal anthers which mature earlier than the central carpel. In time, the style of the carpel will expand and extend the stigma further away from the main body of the flower. When mature, the stigma can be seen to have four lobes.

In Rosebay Willowherb, the later maturation of the female structures certainly encourages cross-pollination. However, as in many (largely) cross-pollinated species, should cross-pollination fail there is a brief period of overlap when the male and female parts are mature at the same time, thereby allowing self-pollination. While cross-pollination is favoured, self-pollination is certainly preferable to no pollination.

Some species such as the common woodland and hedgerow species Lords-and-Ladies (Cuckoo Pint) and the grassland species Cowslip use even more specialised mechanisms; these will be described in later chapters.

Wind and insect pollination

In some self-pollinating flowers, pollen can fall from anthers onto underlying stigmas if the flowers are disturbed or shaken by passing animals or even by the wind. However, most flowers are adapted in further ways to facilitate pollination.

These adaptations usually encourage transfer of pollen by wind (wind pollination) or transfer by insects (insect pollination).

Wind pollination – many people know that Grass pollen can cause hay fever, but not as many extrapolate this knowledge to appreciate that if Grasses produce pollen, then they must be flowering plants. Part of the reason is that Grass flowers are not brightly coloured or showy in the way that most insect pollinated plants are with their brightly coloured petals.

That is not to say that wind pollinated flowers are not adapted: wind pollination can be a risky business, so wind pollinated flowers need to produce vast quantities of pollen as the chance of any one pollen grain reaching a stigma is very small (particularly in cross-pollinating plants). Furthermore, stigmas are often extended to increase the surface area on which the pollen can land – another feature which increases the possibility of successful pollination.

The Grass plant shown in the next photograph shows some

features common in wind pollinated species.

Wind pollinated Grass.
Many elongated yellow anthers, rich with pollen, dangle from the flowers of this plant making it easier for the wind to catch and transfer pollen. This is further aided by the weight of the flowers helping bend the grass stem horizontally, allowing the anthers to be positioned well away from the stem. Less obvious in the photograph are the white feathery stigmas, positioned well above the stamens, thereby reducing the likelihood of self-pollination.

Insect pollination – using insects as a means of transporting pollen certainly increases the chances of successful pollination, but it comes at a cost. Insect pollinated flowers are often large and colourful, with pollen, or more usually sugar-rich nectar, being offered as a reward for visiting insects. Consequently, there is a large investment on behalf of the plant in ensuring that the prospects for pollination are good.

Greater Stitchwort is insect pollinated. It is common in hedgerows and woodland fringes in Spring. As in Lesser

Celandine described earlier, the flowers of Greater Stitchwort are radially symmetrical (circular) and relatively simple in structure. The next photograph, taken in late April, shows five deeply divided white petals immediately above a whorl of shorter green sepals. Ten stamens with bright yellow anthers surround the central carpels.

Greater Stitchwort

Pollinating insects include bees, butterflies, flies and beetles. The simple flower structure of Greater Stitchwort facilitates pollination by a wide range of insects – not a bad thing for plants which flower relatively early in the year when there are fewer insects about. Lesser Celandine and Rosebay Willowherb, previously referred to in this section, can also be pollinated by a range of insect types due to their simple flower structures.

In those species with more complex flowers, the pollination process itself is more complex. Take for example **Gorse**, (perhaps better known as **Whin** in NI), a common plant of poor-quality rocky soils. Due to its thorny stems, it also has a frequent presence in hedgerows as it provides an effective barrier.

Being in flower for much of the year, although usually most vibrant in Spring, Gorse adds colour, even in bleak winter months. A member of the Pea family (scientifically called *Fabaceae*), each flower is surrounded by five sepals although these are fused into an upper and a lower 'case' which separate to expose the golden yellow petals. Each flower also has five petals, two of which are joined centrally making a tube within which are the ten stamens and carpel. This tube forms the *keel* and two more petals form the *wings* at the side of the keel, with the fifth petal as a top cover (the *standard*).

The stamens are held under pressure within the keel, so when a bee arrives searching for nectar, the weight of the bee causes the keel to split, and the sudden release of pressure causes the stamens to spring up shooting pollen against the undersurface of the bee's body.

In the flower in the centre of the photograph, the keel has already split showing the 'sprung' stamens.

Gorse shows a general trend in the flowers of insect pollinated plants with the more complex types being bilaterally symmetrical (symmetrical in two planes only) rather than radially symmetrical (symmetrical in any vertical plane). Note also how the number of sepals and petals are the same with a similar number (or multiple) number of stamens – this is a consistent theme throughout flowering plants.

This book contains examples of even more complex flower structure, with the Orchids perhaps being at the top of this hierarchy. More on Orchids later.

Flowers may occur individually, with one flower per flower stalk as in Lesser Celandine, or in inflorescences (groups) ranging from a small number up to hundreds on one stalk.

Some inflorescences are arranged as spherical (rounded) flowerheads such as is common in *Allium* (Onion and Garlic) species. The next photograph shows how the numerous small tubular flowers of **Devil's-bit Scabious**, a species common in grasslands, are packed tightly together to form a rounded ('pincushion') flowerhead.

OPPOSITE: *Gorse – note the hairy dark-yellow sepals, a protective covering for the other more delicate flower parts when in bud. The petals of each flower are organised as a central tubular 'keel', side 'wings' and an upper 'standard' – all of which are clearly visible in the open flower in the centre of the photograph (taken in late March).*

A Six-spot Burnet moth feeding on the nectar of Devil's-bit Scabious in late August – the long proboscis (feeding tube) of the moth can be seen extending down into the base of one of the flowers to suck up nectar. As they feed, the moths brush against the purple-magenta stamens picking up pollen.

Another component of many flowers are **bracts** – these are usually tough, leaf-like structures found at the base of many flowers and inflorescences. Their main function is protection of the flower or inflorescence, although they can be important in attracting pollinators too. The next photograph shows a flowerhead of Devil's-bit Scabious before the flowers have opened.

Devil's-bit Scabious flowerhead photographed in early August.
Note the sharp triangular tips of bracts below individual flowers
which are yet to open, in addition to the longer sharp-ended bracts
extending out radially immediately below the flowerhead.

Rather than the 'pin-cushion' style, inflorescences are more commonly organised in a more linear or spiked pattern. The **Early-purple Orchid** brightens hedgerows, woods and grasslands in April and May with its flowers ranging from a deep pink to purple in colour.

The next photograph, taken in late April, shows a feature commonly displayed by many inflorescences. This being that the flowers open in sequence from the bottom of the inflorescence up towards its tip. The flowers at the base of the inflorescence are open and mature whereas the ones at the top are not yet open. Inflorescences of this type are called **racemes** and having flowers opening at different times means that the overall flowering period is extended, which increases the likelihood of reproduction.

An Early-purple Orchid raceme. In some of the mature flowers near the base of the inflorescence, it is possible to see the tubular spur (containing nectar) extending upwards from the back of each flower.

Another type of flower and flowerhead arrangement is seen in the Carrot family (*Apiaceae*), a family which includes very common species such as Cow Parsley and Hogweed. In these species the flowers are organised in umbels. An **umbel** is an inflorescence in which all the stalks (rays) originate from a single point. Many species in this family have two levels of umbel structure, giving compound umbels. There are terminal umbels which contain the flowers – these are secondary umbels. Each of these secondary (terminal) umbels is connected by a ray to the top of a common stem thus forming a much larger primary umbel which includes all the secondary umbels (and all their flowers) as shown in the next photograph.

A primary umbel in Hemlock Water-dropwort. Note how all the rays originate from the same point at the top of a common stalk. Each group of white flowers is part of a secondary (terminal) umbel – the rays of these are much shorter but can clearly be seen to be originating from the same level at the top of each of the primary rays. Hemlock Water-dropwort is a plant not to get too friendly with – more on that later!

More ... The Apiaceae family used to be called the Umbelliferae family. Not surprising, as umbels can thought of upside-down umbrellas!

Members of the Daisy family (*Asteraceae*) have what are described as composite flowers – unsurprisingly, these plants are often called composites. Daisies and other similar flowers usually have two distinct types of flower structure arranged within a larger flowerhead. Centrally, tubular disc-florets (or flowers) – yellow in Daisies – are surrounded by the more peripheral, more flattened blade or strap-like ray-florets (flowers) – these ray-florets are white in Daisies – refer to the photograph of Daisies in the Introduction. The flowerheads in this family are usually not rounded in 3D (as in Devil's-bit Scabious) but are more flattened.

The whole flowerhead is held together by bracts arranged together in a whorl or rosette forming an underpinning structure called an **involucre**.

Sea Aster is a typical example of a plant with composite flowers, producing flowerheads up to two centimetres in diameter from July to September. Occurring on rocky shores and in saltmarshes, this species can be found around much of the NI coast. Yellow disc-florets are surrounded by ray-florets which can range in colour from mauve through to white or may even be absent in the flowerheads of some plants.

In Sea Aster, the central yellow disc-florets are hermaphroditic, containing both stamens and carpels, whereas the outer ray-florets contain only carpels.

There are many other variations of composite flower structure. While Sea Aster and Daisy have both disc and ray-florets, Thistles have tubular disc-florets only and Dandelion ray-florets only.

Sea Aster flowerhead photographed in late August showing the central yellow disc-florets surrounded by white ray-florets.

Flowerhead of a Dandelion showing ray-florets only. Older and more mature florets are on the periphery with younger florets opening from the centre.

More ... Dandelions are often seen as weeds, but they are a vital source of nectar and pollen for queen bee bumblebees emerging from winter hibernation at a time when there are very few other flowers open.

More ... Dandelions have been used in traditional medicine for centuries. They have been used as a diuretic to stimulate urine production leading to its French name of *Pis-en-lit* (wet-the-bed). It has been used for a wide range of ailments and it has been suggested it has more beta-carotene (a precursor of vitamin A) than carrot. Dandelion is also a popular food supplement and an ingredient in herbal teas.

There are of course many variations on the themes of flower structures described in this chapter. Many of these will be described in the following pages and some of the adaptations which have evolved in species to maximise the prospects of pollination can only be marvelled at.

Wildflowers of the Coast

For a relatively small area, Northern Ireland has a wide range of different plant habitats. These include coastal habitats, woodland, wetland, grassland, heath, bog and many variations of these.

The actual habitat in a particular area depends on many factors, including underlying rock type, drainage, amount of grazing and human land use.

Nowhere in Northern Ireland is further than a short drive from the coast, meaning that most people are familiar with at least some coastal settings. Of course, there are many different types of coasts, ranging from gently sloping sandy beaches backed by sand dunes to towering coastal cliffs inaccessible to all but the most experienced climber. Coastal habitats are often rich in wildflowers due to their unsuitability for agriculture.

Most people are familiar with the inter-tidal zone which stretches from low to high tide lines. On rocky shores, this zone is normally dominated by seaweeds – plant-like organisms not classified as 'true' plants.

As we review some coastal flowering plants over the next few pages it is worth considering an important feature of coastal plants; many coastal species are seldom found anywhere else. Reasons for this will be addressed later. Rather than covering coastal plants as one homogenous group, they will be separated into three broad groups – flowering plants of the upper shore and saltmarshes, then those of grassy banks and sea cliffs, and finally, the species of sand dune systems.

The upper shore and saltmarshes

The upper shore refers to the area around and just beyond the normal high tide mark. This zone can be rocky or covered in shingle or sand and can be inundated by seawater from the highest tides or at least be subjected to high levels of sea spray. Saltmarshes are those usually flat and low-lying areas which are covered by sea water at intervals, and they retain high levels of moisture even after the tide has ebbed. Where the coastal terrain is relatively flat, saltmarshes can extend inland for considerable distances. In each of these zones, plants must be able to survive having their roots covered by salt water at least occasionally and often much more frequently.

And now for a look at some of the plant species found in these habitats.

Sea Aster, a species with daisy-like composite flowers was touched on in the last chapter. It is a characteristic species of saltmarshes and upper rocky shores, flowering from July to September.

Sea Aster (photographed in mid-August) – this species is clearly able to tolerate having its roots in sea water much of the time.

Another species, typically occurring in the rocky crevices of upper rocky shores, is **Rock Samphire**. In NI it is largely restricted to the County Down coast where it produces greenish-white flowers from June to September. As with Sea Aster, the leaves of Rock Samphire are elongated and narrow, a design that offers a degree of protection against strong tides and harsh coastal winds.

Rock Samphire photographed in late August on the Lecale Coast in Co Down – many of the flowers in this photograph have yet to open. Note how the leaves of the Rock Samphire are fleshy and succulent and therefore excellent water storage organs.

More ... Rock Samphire is a delicacy in some up-market restaurants, and of course, is a target for foragers, particularly in those parts of the British Isles such as Devon and Cornwall where it is most common. There is even reference to foraging for Samphire in Shakespeare's *King Lear*.

Sea Beet is typically found around the drift (tide) line of shingle beaches, in saltmarshes, and may even be present on grassy banks slightly further inland. Sea Beet is sprawling rather than upright and has a deep taproot which penetrates far enough into the ground to give the plant an anchor strong enough to survive harsh conditions. In NI, Sea Beet is most likely to be found on the County Down coast and the rockier eastern parts of the North Coast.

Sea Beet – note the thick, dark green, glossy leaves. The flowers are small and insignificant, as they are in many wind pollinated species, and are present between July and September.

Sea Beet belongs to the Goosefoot (*Amaranthaceae*) family, a family which includes other coastal species such as Annual Sea-blite, Oraches and Glassworts.

More ... Sea Beet is the native precursor of beetroot and sugar beet, commercial crops which have been selectively bred from this starting point. It is also closely related to spinach and is popular with foragers seeking a natural alternative. As the previous sentences suggest, both the fleshy roots and leaves are edible.

Common Scurvygrass is a plant of saltmarshes and upper shores and is widespread around the coast of NI in suitable habitats. It has heart shaped leaves and small white (although often tinged mauve-lilac) four-petalled flowers, with a flowering period which runs between March and August. This species is self-pollinated by a range of insects including flies and beetles.

Common Scurvygrass is a member of the large *Brassicaceae* (Cabbage) family. Species in this family have plants with radially symmetrical flowers each having four sepals and four petals arranged in a cross.

Common Scurvygrass photographed late May in a muddy upper shore. Note the numerous small flowers (each approximately one cm in diameter) arranged in dense clusters at the end of stems and branches, with the four petals of each flower arranged in a cross.

More ... Common Scurvygrass is edible raw or cooked. Its name reflects the fact that it is a species rich in vitamin C and can help prevent or help people recover from scurvy, a vitamin C-deficiency disease. In previous centuries, sailors suffering from scurvy after returning from long sea voyages without access to fruit or vegetables rich in vitamin C, would eat Scurvygrass to help them recover.

More ... The Cabbage family has many commercially important and well-known species including cabbage, mustard, rocket, kale, turnip and radish.

Other relatively common upper shore plants belonging to this family include **Sea Rocket**, a species typically occurring on sand or shingle near the drift line. Sea Rocket has pinnately lobed leaves similar in shape to the closely related commercial rocket, but the leaves in the wild form are fleshier, and it has small lilac to lilac-white flowers. Flowering from June to September, its flowers are in terminal clusters at the end of stems.

Sea Rocket photographed in late August in its typical habitat, a sandy shore close to the drift line. Although sand-blasted and having suffered a degree of damage in a very hostile environment for plants, it is possible to see how the pinnately lobed leaves are very similar in shape to the rocket available in salad sections of supermarkets, although Sea Rocket leaves are much thicker and more succulent.

Sea Rocket is an **annual**, meaning that it completes its life cycle within one year. Annuals are typically found in unstable habitats as the degree of disturbance often prevents individual plants surviving very long. On that theme, it is hard to imagine a habitat more unstable than the upper reaches of a sandy shore.

Another member of the Cabbage family found in upper shore habitats is **Sea Radish**, a species related to commercial radish. It is a species of shingle, rocky and sandy upper shores. Sea Radish is a tall, straggly plant which can grow up to one metre or more, flowering from June to August and producing flowers with four yellow petals arranged in a cross, as is typical for this family.

The next photograph shows the relatively common **Sea Campion**, a patch-forming coastal species with white flowers found in saltmarshes and the upper shore.

Sea Campion photographed in early June growing on shingle and loose rocks around the drift line – flowers are normally produced from June to August.

The photograph of a single Sea Campion flower shows how the sepals are fused together to form a calyx-tube. The calyx-tube forms the deep-pink coloured 'bladder', a structure present in other Campions which we will meet later.

The brilliant white petals of Sea Campion contrast with the deep pink, attractively veined sepals fused together into a calyx-tube.

Sea-purslane, a species more associated with warmer climes, is restricted to the more southern County Down shores of NI. Although relatively common on parts of the Lecale Coast, including in Dundrum Bay, and on Strangford Lough shores, it is not found in NI north of Belfast Lough. Sea-purslane is typically found in the upper, drier parts of saltmarshes. Male and female plants are separate and the unshowy small flowers give the clue that this is another wind pollinated species. Sea-purslane flowers in late Summer.

Sea-purslane photographed in July. This image shows the small indistinct creamy-red flowers.

As shown in the next photograph, the leaves of Sea-purslane are broadly oval (ovate) in shape and fleshy.

The 'fleshy' ovate grey-green leaves of Sea-purslane.

Sea-purslane is in the same family (Goosefoot family) as Sea Beet, which was described earlier. As with Sea Beet it has small, indistinct wind pollinated flowers and thick fleshy leaves.

The rounded pink inflorescences of **Thrift** are common on saltmarshes, rocky shores and cliffs all around the NI coast. Thrift has a strong, deep taproot enabling it to cling on in the exposed places where it is found. Perhaps the species most associated with the coast, the numerous flowerheads, each around one cm in diameter, are evident from late Spring through to Summer. Unsurprisingly, given the bright colour of the flowers, it is insect pollinated.

Thrift photographed in late May on an upper rocky shore. The flowers are at their vibrant best at that time.

More ... The old three-penny-piece, a relic from pre-decimalisation days, had a representation of Thrift in its design.

A species which tends to form low-growing mats, rather than more upright patches or clumps, is **Sea-milkwort**. It is found in saltmarshes, shingle and rocky shores and through its prostrate habit and ability to spread laterally can cover quite large sections of ground. Leaves are small and ovate in shape, and as with so many other plants in these saline-rich habitats, they are glossy and succulent.

Flowers originate singly from the base of leaves, are pink and present from May to September. The colour is provided by five sepals, largely fused together to give effectively a five-lobed calyx, with no surrounding petals. Each flower has five stamens. This photograph of Sea-milkwort was taken in late June at the edge of a saltmarsh on the County Down coast.

Sea-milkwort – this plant is more upright than is usual, as it typically forms prostrate mats covering the ground surface. Note how the small pink flowers originate in leaf axils (the upper junctions of leaves and stems).

Sea Sandwort is another mat-forming species, very similar to Sea-milkwort in appearance and frequent in upper shore habitats. During the flowering season, the presence of greenish-white flowers should prevent confusion between the two species.

Sea Arrowgrass is another species of the upper shore and saltmarsh. At first glance quite like the Plantains – more on Plantains later. Wind pollinated with small flowers produced between June and September, this plant has very thin leaves which are very like those of Thrift.

Sea Arrowgrass in June – note how the flowers extend along almost the entire length of the inflorescence stalk. Note also the very thin, almost wiry, leaves which could easily be mistaken for Grass.

The shore and saltmarsh plants described in this section are obviously all adapted to the harsh coastal environment. Strong winds are a given, which can at least partially explain the thin leaves of many species. Most of the species covered so far are **perennials**, meaning that they can live for more than two years.

Given the harsh conditions, it means that those plants which get through the perilous early stages of survival can gain a foothold and survive in situ for several years.

All the species covered in this section survive in salty soils. Ironically, in an environment where so much water is available, many plant adaptations are linked to the plant's ability to either store water (fleshy or succulent leaves) or reduce water loss by evaporation. Reducing water loss by evaporation is typically achieved by either having a glossy (waxy) surface which largely prevents evaporation through the leaf surface, or very narrow leaves which give a reduced surface area across which evaporation can take place. The problem coastal plants have is that most of the available water is salty, and although coastal plants can still absorb water in this environment, they use a large amount of metabolic 'capital' in doing so, which makes water a valuable resource to be conserved where possible. Virtually all the species which can survive in upper shore or saltmarsh habitats have at least one of the features described above to either store water or reduce its loss.

The adaptations that coastal plants require to survive in salty soils allow them to succeed, but also explains why non-coastal plants (without these adaptations) cannot survive in these habitats.

More ... An interesting footnote to this is that some of these coastal plants are beginning to spread inland along the verges of major roads, a consequence of regular salting by gritting lorries creating saline-rich conditions.

There are many other coastal plants which could easily have been included in this section of shore and saltmarsh species. For example, Annual Sea-blite, the Sea Lavenders, Glassworts and the relatively rare Oysterplant are all interesting and attractive species.

Grassy banks, sea cliffs and clifftops

Slightly further back from the shore is a zone which is subject to sea spray but not flooded by sea water. Some of the species already described such as Thrift occur in this zone but, in contrast, there are species found here that are unlikely to occur in places prone to inundation by sea water.

Soils in this zone are shallow and usually low in organic matter so there are few trees, shrubs, or large herbaceous plants such as Nettles, the conditions being too harsh for them to become established.

Spring Squill is a classic species of coastal banks and clifftops, a species which is now relatively rare. At its best in May, Spring Squill can be seen in several parts of the Lecale Coast, on Rathlin Island and other parts of the North Coast where it turns the short turf a delightful shade of blue in those places where it grows in profusion.

Spring Squill on a Lecale Coast clifftop, looking out over the Irish Sea.

The pale blue to violet flowers of Spring Squill can be seen in more detail in the next photograph. The anthers and carpels are a darker shade, making them stand out from the rest of the flower. The long but narrow, fleshy curved leaves of the plant can also be seen. As with the other coastal plants covered so far, Spring Squill is not tolerant of deep shade (unlike its shade-loving relative Bluebell), a fact contributing to its characteristic presence in short coastal tuft.

Spring Squill photographed in late May. Note that each flower has a perianth with six tepals and six stamens, a characteristic feature of the Asparagaceae (Asparagus) family to which it belongs.

More ... Spring Squill is the county flower of County Down.

English Stonecrop, another beautiful plant, forms dense mats covering sections of rock and areas of very shallow soil. The flowers of this species have five white petals, at their best in mid-Summer, contrasting with the reddish tinge of many other parts of the plant.

This species grows best when in positions that are subject to high light levels and rapid drainage, features which explain its presence carpeting rock outcrop sides and tops. The very small (usually 0.5 cm or less) succulent leaves are as much a feature of it growing on rock as it being subjected to sea spray. English Stonecrop is also found in suitable sites further inland.

English Stonecrop photographed in late June. Note the small but highly succulent red leaves below the flowers. Plants are self-pollinated by a range of insects and are in flower from June to September.

Sheep's-bit is an interesting plant found around the coast. Unlike many other coastal species, it is more likely to occur in peaty rather than sandy soils. Furthermore, it also occurs in more inland peaty areas such as the Mourne Mountains and Antrim Plateau. It is a **biennial**, completing its life cycle in two years, having built up enough resources in the first year to enable it to flower between May and August in its second year. Each plant has a rosette of small leaves radiating out from its base.

Sheep's-bit plants with their bright blue flowerheads (photographed late June). The very thin, long, Bluebell-like leaves in the foreground (some starting to yellow) are leaves of Spring Squill. The very short upright rosette leaves of Sheep's-bit can be seen at the base of the flowerhead stalk in the centre of the photograph.

The small pink to pink-purple flowers of **Wild Thyme** make an impressive sight when growing in profusion along clifftops or other suitable habitats where there is short turf on sandy or rocky terrain. Wild Thyme is common in suitable areas near the coast, but also common in parts of west Fermanagh where the shallow soils on limestone provide the rapidly draining conditions the plant requires.

Leaves of Wild Thyme are small (usually less than one cm), as are the flowers, which are organised in dense heads on stems which creep over the ground surface. Wild Thyme's flowering period runs from May to September. The photograph of Wild Thyme was taken on a coastal clifftop in late June.

Wild Thyme in morning sunshine on a coastal clifftop. Although individual flowers are small, their sheer number in a small area creates an impressive display.

More ... Wild Thyme is only faintly aromatic, unlike some of the varieties available in garden centres.

Many people are familiar with Ribwort Plantain, a common species in many different habitats throughout Northern Ireland including gardens and road verges. **Sea Plantain** is closely related, very similar in some respects (for example flower structure), but very different in other ways such as leaf shape. Significantly, it is also very different in its distribution. As its name suggests, Sea Plantain is mainly restricted to coastal regions, being found all around the NI coast.

Sea Plantain photographed on a cliff top in late June. Note the typical Plantain flowers with protruding stamens (in this case yellow). The very thin linear leaves are an adaptation to strong winds and salty soils.

Sand Dunes

Unsurprisingly, sand dunes typically develop behind sandy beaches rather than rocky or shingle shores. Sand blowing inland is an important feature in the origin and maintenance of dune systems. Northern Ireland has several excellent examples of extensive sand dune systems – think of Murlough Bay (County Down), and the North Coast gems of Magilligan, Benone, Portstewart and White Park Bay for starters. Dunes are among the best places to see both relatively rare and some of the most beautiful plants found in NI.

Dune systems are not homogenous – as distance increases from the sea, the dunes become more stable and the sandy soils become less calcareous (rich in lime) as they increase in organic content, changes which inevitably lead to changes in species composition.

Marram grass is the most common plant in young dunes – the dunes closest to the shore. Relatively few species are found here in addition to the ubiquitous Marram, a plant highly specialised for growing in sand. Marram not only thrives here but is crucial to the success of the dunes as it helps bind the sand together.

Marram grass at the edge of a dune system closest to the sea.

Lyme-grass also occurs at the upper parts of many sandy beaches and in dunes. This species has blueish-green leaves, and this itself is enough to allow it to be easily identified. In amongst these dunes or even at their seaward edge, a small number of hardy species such as Sea Holly and Sea Rocket can be found.

The real gems for botanists, though, are the dune slacks, the flatter, low-lying areas between the dunes which are less prone to the rapid drying which affects the dunes. The dune slacks close to the beach may be prone to flooding with sea water occasionally, whereas slacks further inland may flood following heavy rainfall. Dune slacks (and the dunes themselves) close to the beach favour species which prefer calcareous soils – calcareous soils are soils rich in calcium carbonate (lime). Calcium from broken down seashells produces a soil environment which has an overriding effect in determining the species of plants found.

Species such as Lady's Bedstraw and Common Bird's-foot-trefoil are common, although both are also found in dry grasslands further inland. **Common Bird's-foot-trefoil** is also known as 'Eggs and Bacon', and this is due to the yellow and reddish-orange colour of the flowers as they open. A member of the Pea family, flowers are present throughout late Spring and Summer. Very popular with insects due to it being a rich source of nectar, the flowers become less red and a more uniform yellow as the season progresses.

Lady's Bedstraw is in flower from June to September and it is the sheer number of small yellow flowers, each no more than a few millimetres in diameter, which allows the colour yellow to be the dominating feature rather than the architecture of individual flowers. As with other Bedstraws, the very fine leaves (less than 2 mm wide in this species) are in whorls at intervals up the stem. It is a member of the Bedstraw (*Rubiaceae*) family.

Lady's Bedstraw photographed in mid-July.
A very common dune species, the flowers, each with four golden-yellow petals, are at the ends of delicate stems and branches.

OPPOSITE: *Common Bird's-foot-trefoil in full colour. The flowers are very similar to those of Gorse (referred to in the last chapter). Although very different plants in structure and size, it is not difficult to work out that the two species are closely related due to the very similar flower structures. This photograph was taken at the end of May – later in the summer the flowers have a more uniform yellow appearance.*

More ... The name Lady's Bedstraw stems from its use as bedding and as a mattress stuffing (in medieval times) due to its springy, soft texture, added to its pleasant aroma when dried. The aroma is from coumarin which has a vanilla scent. It has also been used to give the distinctive yellow colour to Double Gloucester cheese.

The flowers of **Kidney Vetch** further enhance the association of the colour yellow with sand dunes. By early June, Kidney Vetch can dominate large swathes of many of our dune systems.

Kidney Vetch typically occurs in large groups.

The small flowers of Kidney Vetch are bunched together to form a rather unusual flowerhead. The flowers open in Spring and can persist into Summer. Very popular with bees and moths, this species provides a rich and important source of nectar.

On the dune slacks in some coastal sites, the flowers of Kidney Vetch have a distinctly red tinge, giving them an almost ethereal appearance in sunlight as seen in the next photograph. Their 'woolly' appearance is due to the hairy sepals which lie at a lower level than the yellow petals.

Kidney Vetch in late May sunshine. Note the hairy 'woolly' sepals below the yellow petals. A member of the Pea family, Kidney Vetch has a very similar flower structure to Bird's-foot-trefoil and Gorse. The slightly folded and pointed leaf-like structures radiating out immediately under the inflorescence are bracts rather than leaves.

More … Kidney Vetch is also called 'Lady's Fingers' and has been used as a traditional remedy for kidney disorders as well as a source of yellow dye.

A particularly attractive plant of sand dunes and other suitable sites where there is rapidly draining soil, little shade, and an absence of large fast-growing competitors is **Biting Stonecrop**. A creeping, mat-forming plant which flowers from May to July, this species is self-pollinated with insects being the pollen vector. It is very closely related to English Stonecrop, as seen by the very similar leaf and flower structures. Both belong to the Crassulaceae (Stonecrop) family. Plants in this family normally have flowers which are radially symmetrical, have five petals, the same number of carpels and have succulent leaves.

The bright yellow star-shaped petals of Biting Stonecrop, photographed at the end of June. Note how the number of carpels matches the number of petals. With its small succulent leaves being excellent for water storage, it is a successful coloniser of rocks, walls and other seemingly inhospitable sites.

Yellow of course is not the only flower colour found in dune systems. The beautiful pale pinks of **Common Centaury** are not particularly common, so they stand out when seen. Furthermore, they often occur as isolated plants or in small groups rather than in the large colonies of some of the other species covered so far.

Common Centaury flowers are present between June and September and are formed as dense inflorescences on branching stalks with clusters on both the terminal and side stalks. It is pollinated by a range of insects, in keeping with the relatively simple flower structure. Common Centaury is common in coastal habitats but is also found in suitable inland sites. In addition to the powder-pink flower colour, plants with white flowers are common.

More … In Greek mythology, the centaur Chiron used a concoction made from Common Centaury to treat a wound caused by a poisoned arrow.

Common Centaury with characteristic pale pink five-petalled flowers. Yellow stamens and stigmas are visible protruding above the plane of the petals. The flowers close at night and in harsh weather, an adaptation which provides protection for the more delicate floral components.

Another coastal plant with pink flowers between June and September is **Common Restharrow**. Mainly restricted to a few coastal sites in County Down, it is a species found in sand dunes and calcareous coastal grasslands. Largely sprawling over the ground surface, the leaves are very sticky to the touch.

The very distinctive pink flowers of Common Restharrow, photographed in late July. A member of the Pea family, this relatively rare species is pollinated by bees.

More … An unusual name – it is suggested that the dense root system of Restharrow disrupted the smooth passage of agricultural harrows over the ground. The Irish name is *Fréamhacha tairne*, meaning roots of nails.

Harebells are delicate and very pretty plants largely restricted to dry, free-draining stable habitats such as mature dunes, clifftops and rocky ledges, and the limestone grasslands and hills of west Fermanagh.

The plants can reach a height of around 40 cm and the pale blue bell-shaped flowers are arranged in clusters at the end of stems and their branches. Harebells are pollinated by a wide range of insects and the flowering period runs from June to September.

The pale blue bell-shaped flowers of Harebell with most of the flowers downturned in their typical drooping pattern. Photographed in late July in a trough between mature dunes.

The next photograph shows a close-up of a Harebell flower showing the corolla tube separating into lobes at its distal end and the characteristic three stigmas.

A Harebell flower showing the distal lobes of the corolla.
The delicate Grass in the foreground, Common Bent, is typical of
many of the native Grass species found in sites without agricultural
reseeding and nutrient enhancement.

More … Those with an interest or knowledge of sewing will understand why the names 'Thimbles' and 'Lady's Thimble' have developed as local names for this species – the initial Harebell photograph gives a good clue! The Irish name for this species is *Méaracán gorm*, meaning blue thimble.

When coming upon **Wild Pansy** in sand dunes or coastal grassland, it is not hard to imagine that its appearance is a consequence of garden escape. Surprisingly, it is native in these and other inland habitats with calcareous soils.

A member of the relatively small Violet (*Violaceae*) family which includes Violets and Pansies, Wild Pansy produces distinctively shaped flowers which can come in a range of colours from yellow to various shades of blue. Wild Pansy has a relatively long flowering period, running from April to October.

The next photograph was taken on a coastal grassland in mid-July. It shows how Wild Pansy flowers are bilaterally symmetrical, have two top petals above two smaller wing petals and a large and centrally positioned lower petal.

The distinctive five-petalled flowers of Wild Pansy. Note the small, weakly-toothed leaves. Wild Thyme with its lilac-purple flowers is in the background.

More ... The photograph shows the Maritime sub-species of Wild Pansy, commonly found growing on bare sand and often associated with the disturbed sand outside rabbit burrows.

Orchids, in the eyes of many botanists, are at the pinnacle of the plant kingdom. A combination of their beauty, floral complexity and rareness – at least for some Orchid species – gives them this exalted position. The Orchid flower shows bilateral symmetry, immediately suggesting a more complex pollination process.

Nonetheless, Orchid flowers have evolved from a relatively simple radially symmetrical flower plan with three sepals and three petals. Typically, the sepals have evolved into a 'hood' at the top of the flower which contains the male and female reproductive structures. Two of the petals laterally bracket the central petal which has become modified into a labellum or 'lip', which is essentially a landing platform for insects. Across the different Orchid species, the labellum has a range of shapes, sizes, colours, textures and markings which are usually distinctive to the species. In the Orchid family (*Orchidaceae*), the petals and sepals are often very highly modified, and as the distinction between sepals and petals is not clear, they may be better referred to as tepals.

One of several Orchid species typically occurring in dune systems is the **Pyramidal Orchid**. This species produces flowers which can range in colour from pink to magenta or even a deep purple and they are normally present between June and August. The dense conical inflorescences can have up to 70 flowers each. The outer edge of the labellum has three lobes of approximately similar size. The leaves brush close to the flower stalk and are lanceolate in shape (long and narrow with maximum width close to the base before tapering gradually at their distal end to a tip).

Pyramidal Orchids are pollinated by butterflies and moths attracted by the rich supplies of nectar in downward pointing spurs at the base of each flower. Quite widely dispersed throughout NI, they are found in other suitable sites with calcareous soils such as the limestone soils of west Fermanagh.

Inflorescence of the Pyramidal Orchid, photographed in mid-July. Note the overall conical shape and the magenta-purple colour. Unlike many other Orchid species, the three lobes on the labellum are approximately equal in size and contain no markings. At the base of the labellum in the mature flowers in the lower half of the inflorescence it is possible to see the overarching hood, within which is the entrance to the spur containing the nectar. The plants with yellow flowers (out of focus) in the background are Lady's Bedstraw.

The **Northern Marsh-orchid** is like most Orchids, a visually stunning plant. Only growing to a height of 30-40 cm, this species is scattered across NI, being found in dune slacks, marshes, damp grasslands, and other localities where the soil is not too acidic. The relatively wide, folded leaves without significant spotting is typical of the species as is the compact inflorescence. The rich purple colour of the flowers is typical too, although it can also be deep red, or anything in between. Flowering is in June and July.

Northern Marsh-orchid, photographed in early June in a dune slack between Benone and Magilligan.

The labellum of individual flowers is approximately diamond shaped with noticeable markings in the form of lines and blotches as seen in the next photograph.

A good place to spot the Northern Marsh-orchid is in the dune slacks on the North Coast such as in the Benone region. However, when coming across plants of this species, it can be difficult to tell if it is a 'pure' form of the species, or one of the many hybrids that it is prone to form with other closely related Orchids.

Highly decorative loops and markings on the broadly diamond shaped labellum of this Northern Marsh-orchid.

The **Green-winged Orchid** was touched on in the Introduction. For many, the most beautiful of all the Orchids, with its aura being further enhanced by the fact that it only occurs in a very small area in one site in NI.

Growing to a height of around 30 cm, the relatively short stem is topped by a short spike of flowers which can be anything from purple to pale pink. The labellum is relatively broad and loosely three-lobed. The central lobe is much paler in colour than the lateral lobes and it is decorated with simple spots. A defining feature for this species are the obvious green lines (veins) running across the tepals (sepals) which form the hood above the labellum. Flowering is from late April to early June.

Not surprisingly, there is interest in why this species is only found in one small site, particularly as it is much more frequent in suitable sites in more southern locations in the British Isles.

This suggests that in its NI site, it is right at the most northernly limit of its geographical distribution.

The Green-winged Orchid photographed in its only NI site in late May. Note the flat labellum with purple spots on the large paler central lobe. The diagnostic green veins are clearly seen on the tepals forming the hood of the flower. This plant is of the pale pink variety found at this site; many of the plants have flowers a relatively uniform purple colour.

Not quite as rare as the Green-winged Orchid, but rare enough, and often hard to find is the aptly named **Bee Orchid**. This strange looking plant is found on calcareous grasslands such as dune systems, or in west Fermanagh, but will also turn up in disused quarries and other sites with a history of disturbance.

Flowering is in June and July with each plant only producing a small number of flowers (usually 2-5) on an upright stem which can extend to around 40-50 cm.

The basic Orchid flower structure is modified to an extreme level in the Bee Orchid. The three pink-lilac sepals are not part of a hood structure as in other Orchids covered earlier but are clearly seen as individual oblong structures with the central sepal extending vertically and the two lateral ones extending as horizontal 'wings'. The labellum has a velvety texture and together with its mosaic colouring it resembles the abdomen of a bee.

Bee Orchid photographed in late June. The two lowest flowers on the stem are not quite fully open. The pink-lilac sepal of the flower on the left is in its final vertical position, but the lateral sepals are not yet in their final horizontal positions. In the flower on the left the two yellow pollinia (in Orchids, the stamens are modified to form discrete masses of pollen which are transferred as an entity during pollination, and these are called pollinia) are visible at the top of the velvety labellum. This plant appears to have four flowers, many fewer than the other Orchid species covered. This is probably due to the space each flower requires once fully open and the high resource input required to produce each flower. The yellow plants in the background are Kidney Vetch.

In many parts of Europe, Bee Orchid pollination is aided by a species of bee which attempts to mate with the labellum, mistaking it for a potential partner. However, the species of bee involved is not found in the British Isles – the British Isles is at the northern geographical limit for this Orchid, but crucially, beyond the northern limit for the bee. In British plants the pollinia grow and curve down over the stigma and due to their proximity, the plants are often self-pollinated. Consequently, as in many other species, self-pollination is a secondary strategy which comes into play when cross-pollination fails.

More ... Global warming and climate change will and do affect plant distribution. Regular flooding of coastal areas and changes in sea levels will of course have an impact. However, increasing temperatures are important too. Increasing temperatures and changing weather patterns can make damp habitats drier and dry habitats wetter but can also affect the range of a species. For example, in southern parts of Scotland, the Bee Orchid is at the extreme northern limit of its range – occurring here but historically not in the more northern parts of the country. Evidence suggests that as temperature is increasing, the range of this species is slowly spreading northwards through Scotland.

As alluded to in earlier sections, many of the species found in dunes and dune slacks (particularly those found at greater distances from the sea), do not require the salty soils that the coastal plants described in earlier sections do. If conditions are suitable, such as the presence of well drained soils, high light levels and the absence of fast-growing competitor species, they can be found in sites well away from the coast.

Of course, only a very small number of the species typically present in dunes or dune slacks have been covered here. Many others can be present in large numbers such as Red Clover,

Knapweed, Wild Carrot, and many native Grass species too as well as 'generalists' such as Ribwort Plantain and Dandelion.

Woodland is next on the list. Very different environmentally from coastal habitats, woodland flowering plants are adapted in very different ways to meet the particular challenges they face.

Woodland Wildflowers

Think woodland – think trees. Large areas of land covered by trees are described as woodland and the type and density of tree species present, together with soil type, largely determine which flower species are likely to thrive within a particular woodland.

Northern Ireland has many different types of woodland: broadleaved woodland (where the dominant trees are deciduous), coniferous woodland (where conifers form the dominant species, often with one or a small number of species in its composition), plantation woodlands of various types and a host of other categories.

Woodlands can be further defined by the main species present, so we can have Oak woodlands or Ash woodlands and so on. A woodland can be described as a damp Ash woodland, in which Ash is dominant but the ground is typically very damp or wet. This is not a definitive list and even within a particular woodland there can be a wide range of different habitats such as glades, clearings, woodland edges, wetter and drier parts, and areas where the underlying rock is different, all features which have considerable effect on the species of flowering plants present.

Botanically, deciduous woodlands are the most interesting in terms of the range of flowering plant species they contain and, consequently, most of this section on woodland focuses on them. Nonetheless, native deciduous woodland is a relatively rare habitat in Northern Ireland, usually restricted to steep ground and river valleys and other land not suitable for agriculture.

Unless a woodland is sparsely populated by trees, the ground-level herbaceous plants have two main but distinct strategies to cope with the reduced light intensities of the woodland floor. They are either adapted to live in low light levels and/or they focus their growth into the short window between when it starts to get warm and be bright enough for growth, and before the tree canopy closes to significantly reduce the light levels.

Oak and Ash are native Irish species, and in woods dominated by these (or other similar but non-native species such as Sycamore) there is often a 'classic' understorey Spring flora in which the woodland floor is carpeted by one or more species. These species include Lesser Celandine, Wood Anemone, Bluebell and Wild Garlic, and each of these can produce dense colonies of plants which extend for many hectares in some woods. These species are adapted to exploit the higher light levels before canopy closure.

Lesser Celandine is usually the first of these species to appear, coming into flower as early as late February and remaining in flower until April or May. In suitable woodland conditions where the soil is moist, Lesser Celandine can form extensive colonies as seen in the next photograph.

*Lesser Celandine (photographed towards the end of March)
carpeting the floor of a deciduous woodland – it is a member of the
Buttercup family (Ranunculaceae).*

More ... Lesser Celandine has also been called Pilewort. This comes from the similarity of the species' root tubers to haemorrhoids and their use as a traditional external treatment for the condition.

By the end of March, **Wood Anemone** is coming into flower. This species with beautiful but simple white flowers can also carpet large areas of woodland floor. Wood Anemone is more likely to be found in soils which are slightly drier than those preferred by Lesser Celandine. The much-divided leaves of Wood Anemone make this a species relatively easy to identify even before the flowers open.

Wood Anemone in flower (photographed early April) – note the deeply-divided trifoliate leaves.

The flowers of Wood Anemone tend to reorientate by continually 'following the Sun' throughout bright days when light levels are high.

Wood Anemone flower – technically the white 'petals' are petal-like sepals (perhaps better referred to as tepals) and number between 6–12. Self-pollinated, with pollination by flies or bees. The flowers of Wood Anemone are normally at their best in April.

Like Lesser Celandine, Wood Anemone belongs to the Buttercup family. It is not surprising these two plants are in the same family as their flowers are very similar. Both species have radially symmetrical flowers, similar shaped tepals of variable number, and there are many stamens and carpels in each flower.

Wild Garlic (also called Ramsons) is another common species of the woodland floor in Spring and grows best in moist, but not waterlogged, soil. This species usually flowers in April and May, and it is easy to identify even when not in flower. The long, elliptical leaves are distinctive, as is the garlic aroma, particularly after rain. In suitable conditions Wild Garlic grows very vigorously and can dominate woodland floors to the exclusion of almost everything else. It is a member of the Daffodil (*Amaryllidaceae*) family.

Wild Garlic carpeting a woodland floor in May.

The star-like flowers of Wild Garlic are arranged in a spherical inflorescence at the end of a flower stalk which typically extends to just above the leaves, resulting in an impressive display of white flowers during the flowering period.

More ... The scientific Latin name for Wild Garlic is *Allium ursinum*. Ursinum is derived from the Latin word 'ursus' which translates as bear. Wild Garlic also grows in many of the northern forests of continental Europe and folk tales record how Brown Bears were partial to Wild Garlic bulbs after awakening from hibernation.

Single inflorescence of Wild Garlic – rather than distinct sepals and petals, each flower has six white tepals and six stamens. Pollination is by a wide range of insects.

More ... Wild Garlic is increasingly popular in cooking recipes. Every part of the plant can be eaten but with the bulbs being less accessible to casual foragers, the leaves (particularly when young) are popular in many dishes from pesto to pizza. The young leaves of March and early April give a stronger flavour than the leaves later in the season. Its large colonies and pungent aroma do not make it easily forgettable. Across the British Isles it is thought to have contributed to many placenames including Ramsey, Ramsbottom and Ramsgate.

More ... Wild Garlic and its more widely used commercial cousin possess antibacterial and antifungal chemicals and are often recommended as excellent healthy food components for this reason. Many other species have these properties too – it's hardly surprising that plants have evolved defences against the types of organisms most likely to harm them.

Examination of the flower structure of Lesser Celandine, Wood Anemone and Wild Garlic (and some of the other species described later in this section) shows that the flowers are radially symmetrical, with separate and free petals (or tepals), and the stamens and carpels easily accessible, allowing them to be pollinated by a wide range of insect species. During the time of year when these species are in flower, there are relatively few insect species around, so it is in the plants' interest to be 'generalists' by allowing a range of different insect species access to nectar and/or pollen, and thereby promoting pollination of their flowers.

The **Bluebell** is the iconic British woodland species. More delicate than Wild Garlic, it flowers from mid-April to around the end of May. The flowers are a tubular perianth and hang on one side of the flower stalk as shown in the next photograph. As with Wild Garlic there are not distinct petals and sepals, and in Bluebell the distinctive blue tepals are fused together with their lips curved back. Flowers contain both stamens and carpels and are pollinated by flies and beetles.

Closely related to Spring Squill, a major difference is that in Bluebell the tepals forming the perianth are fused together for most of their length, rather than being separate as in Spring Squill. Species with petals or tepals fused to form a tube or equivalent are more complex and evolutionarily advanced than the species with more open flowers as described earlier.

Bluebell flowers – photographed in early May in a mid-Ulster mixed woodland. Note the absence of distinct sepals and petals (the two blue strap-like structures at the base of each flower stalk are bracts).

Beautiful individually in their own way, Bluebells are at their best when large drifts are seen in those woods in which soils are not too wet and where there is enough Spring light penetrating to ground level, such as in the next photograph.

More ... The introduction of the alien Spanish Bluebell threatens the native species through the formation of hybrids with native plants. It is relatively easy to identify the Spanish Bluebell (or hybrids with Spanish Bluebell features) – they have wider leaves, flowers hang all around the stem rather than on one side only, the 'bell' is more open, and the anthers are blue rather than cream as in the native species.

A classic Bluebell wood in May – Northern Ireland has many woods which appear like this in late Spring.

In suitable conditions, each of the Spring species described earlier can dominate large sections of woodland floor eliminating competition from other plant species – no bad thing from a plant's perspective.

Rapid early growth is a feature of these species, but the ability of each of them to spread laterally is also important. Relatively heavy seeds which don't spread far helps produce dense populations, but vegetative propagation (asexual reproduction) is perhaps even more important in enabling Spring woodland species form dense colonies which outcompete other plants.

The woodland species described are all perennial and overwinter as underground structures (rhizomes, bulbs, and tubers) which also aid lateral spread.

Rhizomes are underground 'stems' growing just below the soil surface which spread horizontally. At intervals, rhizomes produce vertical shoots which form the visible above ground parts. Extension by rhizome is the main reason for lateral spread in Wood Anemone.

More … Rhizomes such as those in Wood Anemone can be very long lasting and can link many 'plants' together. For example, a dense patch of plants one metre square may all be linked by rhizome connections, raising the question as to whether the group consists of one or many plants! Molecular analysis has shown that stands of Bracken (a common fern which also spreads by rhizome) extending across many metres may technically only be a single plant.

Wild Garlic and Bluebell overwinter as **bulbs** and in addition to the 'main' bulb of each plant, they can produce 'daughter' bulbs when in good growing conditions, each of which will develop into another independent plant, thus maintaining a high plant density in favourable conditions. Lesser Celandine plants produce **root tubers** which are like bulbs in enabling lateral spread.

Vegetative reproduction by these methods ensures that the plants are present in high densities thus reducing the growth of potential competitors, but also enables the colonising of fresh sections of woodland floor.

It is important to note that bulbs, rhizomes, tubers, and other similar structures such as corms, not only aid lateral spread but are also important in allowing the plants to survive through the inhospitable winter season, as the above ground parts die back.

All four Spring species described earlier tend not to thrive on acidic soils such as those typically found in coniferous woodland (although the absence of high light intensities in Spring will cause problems too) or occurring on granite, but they can and do occur in hedgerows.

Another species which flowers in Spring (April-May) is **Wood-sorrel**. Wood-sorrels have their own family (Wood-sorrel or *Oxalidaceae* family), making this a much smaller grouping than some of the other families touched on. Typically, Wood-sorrel spreads mainly by rhizome rather than seed but it normally occurs in small patches and never approaches the lateral spread of the species described earlier. A pretty plant, it can be difficult to spot in woodland, particularly if there is a lot of Wood Anemone, as the flowers are similar at first glance.

The white flower and trefoil leaves of Wood-sorrel – note the five lilac-veined white petals. Although difficult to see in this photograph, there are ten stamens and five styles. Single flowers are at the end of each relatively long flower stalk.

In rain and when dark the leaves fold up and the flower closes as a protective measure as shown in the photograph below.

Rain must be on the way – note how the flower closes and the leaves fold up.

Wood-sorrel is more tolerant of the acidic soils and low light levels characteristic of many coniferous woods than the other woodland species described so far in this section.

More ... Wood-sorrel is one of several plants collectively known as 'shamrock', which is worn by many on Saint Patrick's Day. A close relative of Wood-sorrel is grown and sold commercially for that purpose.

A feature of old established woods is the presence of **Primrose**. Many people associate the Primrose more with gardens than woodland, but it is a classic woodland Spring species. Pale yellow flowers are borne singly on long flower stalks and flowering is typically March to May, although can extend beyond this. When undisturbed and not subject to the presence of more dominant woodland floor species, largish clumps can form due to rhizomatous lateral spread.

A clump of Primroses in woodland – the low growing mat-forming plants on either side of the Primroses are Opposite-leaved Golden-saxifrage, a species typically found in the damper areas of deciduous woodland. It is not surprising that these two species grow together as Primrose does best in soil which does not dry out too much in Summer.

Lords-and-Ladies is a common woodland and hedgerow species, particularly in base-rich soils. Base-rich soils are soils rich in alkaline minerals such as calcium. Lords-and-Ladies is common throughout NI apart from upland and mountainous areas such as the Sperrin and Mourne Mountains and parts of the Antrim Plateau.

Arrow-shaped, large leaves emerge in February and March, and flowering is in April and May. Lords-and-Ladies belongs to the Lords-and-Ladies (*Araceae*) family, another family with relatively small numbers of species.

The way in which cross-pollination is encouraged in Lords-and-Ladies is particularly complex and interesting and is all down to the plants' very unusual structure. Individual flowers are unisexual – either male or female. The flowers are arranged in a particular order on a central cylindrical spike called a **spadix**. The spadix has a whorl of female flowers at its base with a whorl of male flowers just above the female flowers – the flowers are small and very simple with no sepals or petals. Above the male flowers is a ring of hairs (sterile flowers) and above this the spadix extends into a purple (occasionally yellow) tip. The spadix is sheathed by a leaf-like bract called a **spathe** as shown in the photograph below.

Lords-and-Ladies showing the purple spadix tip surrounded by the sheathing spathe. The ring of hairs (and the whorls of male and female flowers) is in the section of spathe just below the opening. (Photographed in late April)

But what is the link between the strange arrangement of the flowering apparatus and pollination? The spadix emits an aroma (unpleasant to humans) when the female flowers are fertile which attracts small flies to the opening of the spathe. The flies bypass the ring of hairs just below the opening of the spathe and travel down into the enclosed section but due to the orientation of the hairs they are trapped and unable to escape back out. If the flies have picked up pollen from another plant some of this will be used to cross-pollinate the receptive female flowers. Successful cross-pollination brings about several changes within the spathe – the male flowers on the spadix start releasing pollen and the ring of hairs trapping the flies rapidly withers, thus allowing the flies to pick up pollen as they escape from the enclosed section of spathe. These flies (armed with pollen) are now equipped to pollinate the next Lords-and-Ladies plant they are attracted to.

Following successful pollination and fertilisation, the characteristic orange-red fruits of the species appear in July and August, well after the rest of the spathe and spadix has withered.

The very visible but poisonous orange-red fruits of Lords-and-Ladies, photographed in late August.

More ... Lords-and-Ladies is a species which has gained many names over time. Cuckoo pint is probably the most common alternative name. Parson-in-the-Pulpit is another version, but many of its names are linked to the similarity between the shapes of the spadix and the spathe opening to parts of human genitalia. From this perspective, it is not hard to deduce the origins of the names 'Adam and Eve' and 'Bulls and Cows'. There are of course other 'X-rated' alternatives!

As an alternative to concentrating growth into the Spring season in woodland, some species are adapted to grow in woodland margins, gaps and glades, or any part of the woodland which allows enough light to penetrate through the year. These places have higher light levels than areas with a complete tree canopy coverage, but there is enough shade to reduce water loss by evaporation compared to fully open sites. Furthermore, there is less competition from vigorous Grasses which would be expected outside woodland. A species which typically occurs in these brighter sites is Red Campion.

Red Campion is a member of the *Caryophyllaceae* (Pink) family. Plants are separately sexed. Red Campion has a particularly long flowering season, with its attractive flowers present from March right through to October or November. The photograph opposite, taken in mid-June, shows the reddish-pink flowers of Red Campion.

The side view of Red Campion in the next photograph shows flowers of this plant opening in mid-April.

These flowers are from a male Red Campion plant as the protruding stamens testify. The five petals of Red Campion are so deeply divided that it gives the impression that each flower has ten petals.

Note how the corolla of Red Campion is fused for about half its length before the individual petals separate and fan out. Closely related to Sea Campion, note the fused calyx-tube, although much less inflated in this species compared to its coastal relative.

Wood Speedwell, which is a bit more tolerant of lower light levels than Red Campion, is a species which tends to sprawl over the surface of damp woods. Although found in suitable woods throughout most of NI, it tends to grow in relatively small clumps rather than having the extensive spread of many other woodland plants. The pale lilac flowers of Wood Speedwell appear from April to July.

Wood Speedwell showing the characteristic lilac flowers with purple lines.

Some species stand out more for their leaf features rather than their flowers. One such species is **Woodruff**. This species has between six to nine (but more frequently eight or nine), broadly similar elliptical-lanceolate leaves arranged in a series of whorls up their unbranched stems. The small star-shaped white flowers with four petals appear during May and June. A patch-forming species of damp shady places, it can most easily be spotted in mixed deciduous woodlands but can also appear in hedgerows.

Patch-forming Woodruff, photographed at the start of June. Note how the leaves are in whorls of eight-nine leaves per whorl. The small white flowers (present from May-June) each have four petals.

Woodruff is a member of the Bedstraw family. Although not visible in the photograph of Lady's Bedstraw in an earlier section, this closely related species also has flowers in the shape of a cross. The leaves and flowers of Lady's Bedstraw are much smaller than those of Woodruff.

More ... The Latin scientific name for Woodruff is *Galium odoratum*. The '*odoratum*' refers to the sweet scent of the plant when dried. It is rich in coumarin, the same chemical which gives newly mown hay its distinctive smell.

More ... Although members of the Bedstraw family can have flowers with four petals arranged in a cross, they are not easily confused with members of the Cabbage family which also have flowers with petals arranged in a cross.

Bugle occurs in damp woodlands from April to July. This species spreads rapidly with its overground runners rooting at intervals to produce new plants. A member of the large Deadnettle (*Lamiaceae*) family, it has the characteristic square stems and leaves arranged in opposite pairs of that family.

Bugle photographed at the end of May, by which time the tree canopy had closed. Many of the other smaller (non-flowering) plants visible in the photograph are young Bugle plants forming on runners which can be seen spreading across the woodland floor. A single Bugle inflorescence is shown in more detail in the chapter on grasslands.

More … Unlike rhizomes, runners spread along the ground surface and are not as long lasting. In general, runners can spread faster than rhizomes as there is less resistance to growth along the ground surface than there is through the soil. Additionally, being much shorter-lasting, runners usually do not have an overwintering or long-term food storage role.

Wood Avens, perhaps better known as Herb Bennet in some areas, is in flower from May to the end of the Summer. Fairly common, it is not too choosy about its environmental requirements, being able to grow in both low and dense shade and in a wide range of soil types. The yellow petals of Wood Avens are widely spaced to the extent that the sharply pointed sepals are visible between them. Although it has five petals and five sepals, it has numerous stamens and carpels. The flowers and flower stems characteristically droop as they mature, thus protecting the more delicate floral structures from the elements.

Wood Avens with its widely spaced petals allowing the stout and sharply tipped sepals to be clearly seen.

Herb-Robert is a very common plant across a range of settings in NI, being found in hedgerows, shingle and rocks at the coast. It is also a common woodland plant occurring at path edges and more open areas where there is some shade, but also a reasonable amount of light getting through the canopy. It is a member of the *Geraniaceae* (Crane's-bill family). Flowering from April to October, Herb-Robert is still in flower at a time when very few other species are.

Although low growing and often scrambling across the ground and through other plants, Herb-Robert is always worth a second look. The flowers are small, often less than 15 mm in diameter but they have five pale pink or pink-lilac petals with orange anthers symmetrically positioned around the yellow stigmas. At the base of each leaf is a nectary which helps attract pollinators. As with many of the woodland species covered, Herb-Robert has a simple open flower which enables pollination to be completed by a range of insect species.

Herb-Robert growing beside a woodland path and photographed in late May. Note the radially symmetrical open flowers. This photograph also shows the deeply dissected leaves of this species.

Flowering plants of the darkest woodland recesses

And now for two very unusual plants – or unusual if you expect all plants to have green leaves. In the deepest shade, such as where the tree canopy fully closes to allow very little light through, very few herbaceous flowering species are evident. With very low levels of light penetrating to the ground layer, photosynthesis can be difficult or impossible, so a very small number of flowering plants adopt other strategies. This includes adopting a saprophytic type of nutrition; **saprophytes** break down dead organic matter, such as fallen leaves and dead wood, and then absorb and utilise their nutrients.

The most obvious saprophytes in woodlands are fungi – this explains why woodlands are a great place for fungal forays – there are just so many there.

But very strangely, some plant species are saprophytes too. The rare **Bird's-nest Orchid** is an example of a species which feeds saprophytically. Very unusually for plants, this species has no chlorophyll and therefore has no green colouring. But of course, it doesn't photosynthesise so doesn't need to be green.

The Bird's-nest Orchid's favourite habitat is in the deepest woodland recesses where the light levels are very low. It is typically found under Beech where there is the characteristic build-up of leaf litter and little competition from other ground-level plants. Additionally, it prefers non-acidic soils such as those present in the woodlands of western Fermanagh – this also explains why this species is not found in coniferous woods where, although the light levels may be suitable, the more acidic soils associated with coniferous woods are not.

Honey brown in colour, Bird's-nest Orchids do not possess true leaves, but do have leaf-like modified scales on their inflorescence stalks. In terms of accessing nutrients, the business part of the plant is below ground level, with the only above-ground part being the inflorescence stalk and flowers which appear from late April to June. In addition to producing seeds, the species can spread by rhizomes, but it is not very prolific with there usually being only small numbers of plants in any one place in any year.

Hard to spot against the leaf litter – the Bird's-nest Orchid.

More ... The 'Bird's-nest' part of the name is due to the roots of the plant being tangled to the extent that they resemble a (very) untidy bird's nest.

The flowers are arranged in a raceme inflorescence containing around 50 individual flowers. Close examination of opened individual flowers in the next photograph shows that the labellum or 'lip' tepal is forked and divided with the two tips diverging; the other tepals in each flower form the overarching hood, within which is the pollen. Pollination is often by flies and beetles.

Inflorescence of the Bird's-nest Orchid – note how the lower flowers have opened first. Note also how the labellum (lip) splits to form two diverging lobes. The creamy objects seen within the overarching hood of opened flowers are the pollinia (masses of pollen).

More ... Saprophytes have a very important role in woodland ecosystems. The large quantity of dead wood produced in these habitats is broken down by fungi and other saprophytes. Apart from their role in decomposition, fungi have other roles which impact on the life of trees and herbaceous plants. Many trees and plants have fine fungal threads called hyphae extending between the cells of their roots. These microscopic fungal threads grow throughout the soil, and in effect, extend the area across which the plant roots actively operate. The trees and plants benefit as their association with fungi increases the uptake of water and nutrients from the soil. It is a two-way relationship in that the fungi gain sugars and other compounds manufactured by the plants. This relationship between plant and fungi can be so important that some species of trees and plants, including many Orchids, cannot survive without this relationship.

More ... Recent research has shown that these fungal threads link many of the trees in a woodland together, allowing a degree of previously unknown communication. Nutrients can pass along this network from tree to tree, supporting those under stress or younger saplings. Additionally, it appears that if one tree in the network comes under significant insect attack, neighbouring trees can be 'alerted' by chemical communication. This allows those trees in the vicinity to rapidly build up their chemical defences against such attack. Unsurprisingly, this underground network involving trees and fungi has been described as a 'wood-wide-web'.

The cream-coloured **Toothwort** is another unusual plant which lacks chlorophyll. As with the Bird's-nest Orchid, it requires strategies other than photosynthesis to gain its nutrients.

However, there are subtle differences between the modus operandi of the Toothwort and the Bird's-nest Orchid. The Orchid uses nutrients which have been released from the breakdown of dead organic matter, whereas the Toothwort proactively extracts nutrients from the roots of living trees, typically those of Hazel, Cherry Laurel and Elm.

Toothwort does this by using special structures which penetrate the roots of its host trees and taps into the nutrients in their cells. This takes the Toothwort a stage further and classifies it as a **parasite**. Only the flowers and a few scale-leaves are above ground in Toothwort, with most of the plant existing as an underground rhizome. Toothwort is a member of the *Orobanchaceae* (Broomrape) family, a family in which all members are parasitic or at least semi-parasitic on other plant species.

Toothwort can be found in woodland in April and May – it is not as rare as the Bird's-nest Orchid but, nonetheless, not common.

Toothwort photographed in early April – note the creamy-pink colour and the flowers organised as a one-sided spike. The styles and stigmas of the flowers are clearly visible and extend well beyond the rim of the corollas.

The woodland species described so far are all species typically occurring in mixed deciduous woodland such that found under native Oak and Ash, with an understorey of smaller species such as Hazel. Sycamore is often found in mixed deciduous woodlands too. Beech gives a denser shade and is more likely to provide a potential habitat for those species adapted to very low light levels such as the Bird's-nest Orchid.

Many of the herbaceous species listed such as Wood Anemone, Wild Garlic, Wood-sorrel, Woodruff and Bird's-nest Orchid are indicators of very old or 'Ancient' woodland, woodland which has been in existence for centuries. These and other similar species are long-lived perennials adapted to live in the relatively stable woodland environment. However, they are not good at colonising distant sites. Consequently, freshly planted woodlands can take a very long time to acquire a rich woodland ground flora.

Mixed deciduous woodlands are typical in lowland sites where the soil is relatively rich and alkaline or neutral, or at least not too acidic. Other ground conditions due to differences in underlying rock type and drainage favour other combinations.

In wetter areas, such as in low-lying areas prone to flooding and lake fringes, species such as Alder, Downy Birch and Willow predominate. These species are common in many of the low-lying areas on the fringes of Lough Neagh. In these woods, the familiar woodland flora is typically replaced by wetland species such as Meadowsweet, Yellow Iris and Water Mint, species which are covered in the next chapter on wetlands.

In the damp limestone river valleys of Fermanagh, Ash with an understorey of Hazel is frequent. Woods in which Ash is dominant often have extensive drifts of Spring species, partially because Ash comes into leaf later than other species, but also because the pinnate leaf arrangement in Ash allows more light penetration than the solid entire leaves of Sycamore, Oak and Beech.

In the same way that the tree species present in a wood

can give clues about the type of underlying rock, the drainage, whether the soil is acidic, alkaline, or close to neutral and the microclimate, so do the herbaceous plant species. These are often a better indicator as they have shorter life cycles and therefore their densities and distribution typically respond faster to change.

Coniferous woodlands do deserve a mention. In Northern Ireland, commercial plantations have tended to consist of uniform stands of one or a few conifer species such as Sitka spruce. Being evergreen, ground-level light intensities remain low all year so there is no window of opportunity for Spring species such as Lesser Celandine and Wood Anemone. Furthermore, the acidic soils which tend to develop under conifer are not conducive to the growth of many species of flowering plants.

The photograph below shows the paucity of ground-layer flora under a stand of **Scots Pine**.

Scots Pine and very little else!

Ferns and mosses can thrive on the floor of coniferous forests and some flowering plants of course do. Wood-sorrel, already referred to in this woodland chapter does well in coniferous woodland, as does the Foxglove in those places where there the light intensity is higher such as in firebreaks and plantation woodland fringes.

Wildflowers of Wetlands

In Northern Ireland, we are never too far away from the coast, but most of us are even closer to a habitat which could be described as 'wetland'. For our purposes, wetland habitats include lakes, ponds, rivers, canals, and their margins, in addition to marshes, wet meadows and other similar habitats.

Species able to thrive in these habitats need to be able to grow in water, or in soils that are frequently or always waterlogged, or at least very wet. Consequently, many of the species found are specialists, adapted for wetland habitats to the extent that they struggle to survive in drier environments.

Water bodies such as lakes, ponds, rivers, canals, and their immediate margins are an obvious place to start.

Lakes, ponds, rivers, canals, and their margins

Duckweeds are true free-floating flowering aquatic plants which float on the surface of ponds, ditches, disused canals and other water bodies where the movement of water is not too rapid or disruptive. Typically, a population of Duckweed contains several very similar Duckweed species of which Common Duckweed is often the most common.

Duckweed photographed in early July. In stagnant or slow-moving water, Duckweed can cover much of the water surface, particularly if the water is enriched with nutrients. Individual plants usually have 1–4 small leaves floating on the surface with short, fine roots extending down into the water.

Water-lilies are adapted to grow in disused canals, large ponds and lakes in places that are not too deep and where there is no flow of water, or where the flow is relatively gentle. Unlike Duckweed, Water-lilies are attached to the mud at the bottom and have long stems reaching up to the leaves floating on the water surface.

Yellow Water-lily is relatively common in suitable habitats. In comparison to the leaves, the flowers are small (usually less than six cm in diameter) and have yellow petals surrounded by yellow-green sepals. The floating oval leaves are sizeable, being up to 40 cm or more in length.

Yellow Water-lily photographed in mid-July (flowering period is June-September). Note the large floating leaves and the much smaller yellow flowers held on stout stalks well above the water surface. Between the Water-lily leaves most of the available water surface is covered by Duckweed.

More ... Yellow Water-lily is also known as 'Brandy-bottle', largely due to the brandy bottle shape of the seed pods and the alcohol smell of the flowers. From an evolutionary perspective, Water-lilies are closely related to the earliest flowering plants.

White Water-lily is rarer than its yellow relative and is much less likely to be found in nutrient-enriched waterways. White Water-lily is a plant of nutrient-poor lakes. Its flowers are much larger, being up to 20 cm in diameter with white petals and green sepals. It has a shorter flowering period than Yellow Water-lily, with flowering concentrated around mid-Summer. White Water-lily has more rounded leaves, usually less than 30 cm in diameter.

White Water-lily photographed in mid-July.

Both Water-lily species described above belong to the *Nymphaeaceae* (Water-lily) family.

Emergent flora is a term which describes those plant species growing at the water's edge and consequently they have the lower sections of their stems and their roots immersed in water.

Yellow Iris (also known as Yellow Flag) is a well-known and common example of the emergent flora in NI. This species can be found in sizeable colonies largely due to lateral spread by rhizome. The narrow sabre-shaped leaves can be a metre or longer in length as can the flower stalks which support large bright yellow flowers.

The complex, bilaterally symmetrical flowers are present in mid-Summer and are pollinated by bees. The shape, size and colour of Yellow Iris flowers makes this species easily recognisable when in flower.

The unmistakeable flower of Yellow Iris – note the darker orange markings on the lowest tepal – these act as guidelines for visiting bees seeking nectar.

Yellow Iris is a member of the *Iridaceae* (Iris) family, which includes Crocuses and Gladioluses.

More … The Yellow Iris flower is thought to be the inspiration for the fleur-de-lis symbol, a symbol which although more widely significant in France, has also played an important role in British culture, including being the Scout symbol and is also used in heraldry and has even appeared on historic British coins.

Another member of the emergent flora is **Bulrush** (Great reedmace). Bulrush plants are very tall, extending up to two or more metres above water level. In suitable habitats, such as the fringes of lakes and large ponds, it can be present in large numbers producing dense stands.

The cylindrical Bulrush flowerheads are present between June and August. The larger chocolate-brown female flowers are sited immediately below the more distal creamy-yellow male flowers as shown in the next photograph.

Bulrush as part of the emergent flora at a lake edge, photographed in late July. Note the terminal flowerhead with a mass of small creamy-coloured male flowers above and continuous with the darker brown female flowers which lie below the male flowers.

Many lakes and other waterways are fringed by **Common Reed** – another regular component of the emergent vegetation zone. Spreading by rhizome, Common Reed can also form very dense and large stands. Its broad leaves can be up to two cm in width. Common Reed is also a common species in marshy ground, which can be a significant distance away from a larger water body.

Large feathery and dark-chocolate to purple coloured flower spikes are characteristic of Common Reed. Flowering from August-October (this photograph taken late August), its small flowers are wind pollinated.

With its appearance that of a large overgrown grass, it is not surprising that Common Reed is a member of the very large *Poaceae* (Grass) family.

More ... Common Reed is a very versatile plant, being used by humans in many ways. It has been used for thatching roofs for a very long time. More recently, it has been used in reed-bed systems for purifying wastewater in isolated houses and farms where mains sewage is not available. It is also used for fodder for livestock.

Much rarer is the beautiful **Flowering-rush**, a species mainly occurring as part of the emergent flora of Lough Neagh and Lough Erne, but also found in the rivers which feed and drain these large lakes. As with the other emergent species covered, Flowering-rush has very long (up to 1.5 m) simple leaves, which emerge from the base of the stem, and the rhizomes which facilitate the building of dense stands due to lateral spread.

Flowers are arranged in a cluster in a terminal flowerhead. Each flower has three pink sepals and three pink petals (together forming a perianth of six tepals), enclosing the stamens and carpels. The flowering period is July to August.

Flowering-rush photographed in early August. The flowers can be seen to be organised as a terminal cluster, with individual flowers having pink tepals surrounding stamens with yellow anthers and dark red-purple carpels.

More ... Flowering-rush is not a Rush! It is the only member of Flowering-rush (*Butomaceae*) family of flowering plants.

As outlined earlier, the emergent flora is often present as dense stands. The species adapted for this zone multiply rapidly and dominate potential competitors. The zone immediately inland from the waterway edge is also often densely populated by another group of plants, species which while normally not found growing in standing water, are adapted to being inundated occasionally and can thrive in very wet soils.

Due to slightly different environmental requirements, a transition of species can often be seen at lake and river edges, as shown in the next photograph.

Note the zonation at this lake edge. Bulrush at the water edge, then Meadowsweet (with the frothy cream-coloured flowers) and then further inland a zone of Great Willowherb (with the pink flowers). The lake itself is densely populated with Yellow Water-lily, suggesting significant nutrient enrichment of the water.

Plants of marshes and very wet soils

Marsh is formed on mineral ('normal') very wet or waterlogged soils but not peaty soils. Very wet or waterlogged habitats on (acidic) peaty soils are described as bog (bog will be covered in a later chapter). The plants covered in this section are those adapted to live in waterlogged or very wet soils. The two non-emergent species identified in the previous photograph (Meadowsweet and Great Willowherb) fit neatly into this category.

Meadowsweet is very common throughout NI. A member of the Rose (*Rosaceae*) family, these plants can be found in any suitable habitat where the soil is wet enough. Meadowsweet can form large dense clumps up to one metre in height and spreads vegetatively by rhizome.

The flowering period is June to September and the frothy clusters of creamy-white flowers are a common sight on lake margins and riverbanks. Pollination is by insects.

Meadowsweet flowers photographed at the end of August. Note the prominent stamens and the 5-6 small white petals on each flower. Also note that the flowers are in branched racemes (a branched raceme is called a panicle – in a panicle, each branch behaves as a raceme). Characteristically, the inflorescence stalks have a reddish tinge.

More … Named Meadowsweet as it is sweet smelling and commonly found in (damp) meadows.

Great Willowherb is another very common herb in marshes and other similar wet habitats, usually present in large stands with individual plants growing to 1.5 m. Again, vegetative spread is by rhizome. This species flowers in July and August with each flower having four sepals and four purple to pink (occasionally white) petals. Pollination is by bees.

Great Willowherb photographed in mid-July. Note the very prominent four-lobed white stigma with the tips arching back. There are eight stamens with the four outer ones being much longer than the four positioned more centrally.

Another species of riverbanks and lake edges is **Yellow Loosestrife**. A member of the *Primulaceae* (Primrose) family, it can grow to a height of 1.5 m, but is typically less than this. Spear-shaped leaves are arranged in whorls of 3-4 up the stem.

Flowers are arranged in dense terminal clusters with the long stalks of each cluster originating in leaf axils. The brilliant yellow corollas have five petals joined at their bases but separate at the distal end to give five lobes. The petals are often orange at their bases. Flowering is during late June, July and August. As with so many similar species, Yellow Loosestrife can spread vegetatively by rhizome, in addition to sexually by seed.

Yellow Loosestrife prefers non-acidic soils in shady lake and riverbank habitats, being relatively common on the shores of Lough Neagh and Lough Erne and the rivers which flow into and out of these lakes.

Yellow Loosestrife photographed in late July on a shaded riverbank. Note the leaves in whorls of 3-4 at intervals along the stem. The bright yellow flowers are arranged in dense clusters on relatively long stalks originating in leaf axils.

More ... Yellow Loosestrife is a popular garden plant so it can be difficult to determine if a stand is native or a consequence of garden escape.

Coming across **Marsh-marigold**, it is not difficult to deduce that it is a member of the Buttercup family. Also called Kingcup, due to the large cup shaped flowers, Marsh-marigold brightens up many stream edges and ditches in Spring and is found throughout NI. Clusters of flowers are supported by relatively thick and strong stems.

The flowers have golden-yellow, petal-like tepals (technically sepals) as seen in the next photograph. Pollination is by a range of insects.

Marsh-marigold in flower at a streamside in early May. Note the clusters of golden-yellow flowers on stout stems.

More ... The Irish name for Marsh-marigold is *Lus buí Bealtaine*, the 'Yellow Shrub of May'.

While Marsh-marigold has classic 'Buttercup' flowers, **Common Fleabane** has the classic composite flowerheads of the Daisy family. Unlike Sea Aster and the common Daisy, in Common Fleabane the disc and ray-florets are the same yellow colour, although the more central disc-florets are a deeper shade of yellow. Common Fleabane often grows in dense clumps, spreading vegetatively by rhizome and typically grows to around 50 cm.

The flowerheads in Common Fleabane are up to three cm in diameter. Flowering is from July to September – this photograph was taken in late August.

More ... Common Fleabane was once believed to drive out fleas and for this reason used to be used in bedding.

Another species of wet marshy ground is **Marsh Woundwort**. Typically growing in dense clumps or stands, Marsh Woundwort grows to a height of around one metre or a bit less and is common in suitable habitats.

Relatively complex flowers are arranged in whorls and are present between July and September. The pink-purple corollas have a hooded upper lip within which the stamens sit. The flatter lower lip has white markings as shown in the next photograph.

Note the purple flowers with upper and lower lips and the many insect guide markings. The square stem and leaves in opposite pairs of Marsh Woundwort are characteristic of the Dead-nettle family.

Another member of the Dead-nettle family, common on stream edges and other damp habitats is **Water Mint**. As with Marsh Woundwort, it has a square stem and broadly oval leaves tapering more gently at their distal ends, organised in opposite pairs with each pair at right angles to the leaves above and below. Stems and leaves often have a purple tinge.

Water Mint photographed in late August. Note the terminal flowerheads.

A relatively late flowering species, Water Mint flowers from July-September. An inflorescence is shown in detail in the next photograph.

Water Mint flowerhead, cylindrical with a hemispherical top. Note the small individual flowers with pale lilac corollas. The petals are fused for much of their length leading to a four-lobed outer rim. Stamens with magenta anthers extend well beyond the corollas.

More … The scientific name for Water Mint is *Mentha aquatica*. In this example, it is easy to see the link between the common name and the scientific name. This is not always the case of course!

We have come across relatively few members of the Carrot family so far. Rock Samphire, covered in the section on the upper shore and saltmarshes is a member of this family. A key characteristic of this family is the organisation of flowers as umbels as discussed in an earlier chapter.

Wild Angelica, a common plant of marshes, ditches and waterway margins, is a typical member of this family. Plants can grow to two metres or more but are typically shorter. This species has purplish stems as seen in the next photograph. The large sheaths visible at the junctions of side branches serve as protective coverings for flowerheads before they expand and open.

Although the off-white umbels can be up to 15 cm in diameter, the individual flowers are only around two mm each. Flowering is from July to September. This photograph of Wild Angelica was taken in early August.

Wild Angelica showing the characteristic purple flowerhead stalks and protective sheaths at branch origins.

More ... The Carrot family contains many species which are poisonous if eaten. Perhaps the most poisonous is Hemlock Water-dropwort, a species surprisingly common in damp habitats. Nonetheless, it also contains species well known as vegetables including carrot, parsley, parsnip and celery.

Some marsh plants are equally at home in damp woodland, showing they can cope with significant shade in addition to high moisture content. One such example is **Water Avens** – our second example of a plant in the Rose family in this section. Growing to a height of up to 50 cm, flowers appear between May and September. As with many other species, the flowers droop downwards when mature leading to the yellow stamens hanging below the corolla lip.

The flowers of Water Avens are an uncommon colour in nature – apricot (dusty pink) petals being surrounded by purple sepals. Pollination is by bees.

More … Many species of plants hybridise naturally with closely related species – unlike in the animal kingdom, where hybridisation is much rarer. When growing in close proximity, Water Avens hybridises with its near relative Wood Avens. Hybrids tend to have flowers intermediate in colour between the two parental species.

Marshes of course are heterogenous; at one extreme they are almost indistinguishable from the edge of a water body such as a lake, whereas at the other extreme it can be difficult to pin down where a marsh morphs into a damp meadow. Several of the plants covered so far in this section can also be found in damp meadows.

A plant in this category is **Ragged-Robin**. Plants can grow to around 75 cm in height and Ragged-Robin is a species unlikely to misidentified when in flower. Flowers are loosely clustered at the ends of branches. The petals are a deep rose-red, but it is the way they are deeply dissected which makes them stand out. Pollination is by butterflies and long-tongued bees which feed on nectar produced deep in the tubes formed by the lower sections of the petals. Flowering period is May to August.

OPPOSITE: *A flower of Water Avens showing its classic 'nodding' orientation (photograph taken early June). Note the colours of the different floral components. The photograph also shows a 'flower' which is now at the fruiting stage. Individual fruits are hairy and have long hooked styles which help in seed dispersal if snagged on the fur of passing animals.*

Ragged-Robin photographed in a 'rushy' pasture in late June. Note the very deeply divided, deep pink petals. Closely related to the Campions, and similarly classified as being in the Pink family, the sepals of this species are also fused to form a calyx-tube which surrounds the lower part of the petals.

Plants tend to be adapted for the environments within which they are normally found. Species typical of wetland habitats clearly are adapted to thrive, or at least survive, in these environments.

For example, Water-lily species have leaves which float on the water surface and they are adapted to do this. Water-lily leaves are large and broad which aids flotation but, additionally, they have large air spaces within the leaves which further aid buoyancy. Most leaves have minute pores (stomata) on their surface which allows the gases required for photosynthesis and respiration to enter (and the waste gases to leave). In most plant species these pores are mainly on the lower surface of the leaf. However, in Water-lily leaves they are on the upper surface as the lower surface is under water.

Many wetland species have hollow stems. This reduces the metabolic demands of the plant as there is less tissue to service without having a significant effect on stem strength. Crucially though, this design allows oxygen to diffuse down the stem to the roots – roots which are often under water and therefore in an environment with a much lower concentration of available oxygen.

Apart from (usually minor) changing water levels, most wetland habitats are normally quite stable. Stable habitats favour perennial plants adapted for a particular environment as they can become established and dominate the habitat to which they are best suited. Many of the species described in this section have underground rhizomes, structures which enable the plants to survive the harsh winters and allows them to spread vegetatively and dominate the available ground.

Obviously, only a small number of the species present in wetland habitats have been covered in this section; nonetheless, the species described give a good flavour of those likely to be found.

Not so welcome – poisonous plants and invasive species

Toxic and poisonous plants

Some wetland wildflower species in NI, native or not, are harmful to humans and other animals and/or are harmful to the environment. Harmful plants are not restricted to wetland habitats of course.

Many plant species possess toxins which harm animals. This is not surprising as the presence of toxins can deter potential herbivores from using the plant for food.

One plant which takes this to extreme is **Hemlock Water-**

dropwort, regarded as the most poisonous plant species in the British Isles. A member of the Carrot family, Hemlock Water-dropwort is surprisingly common in wetland areas such as lake fringes, river and stream edges and marshy areas in general. A perennial plant which grows to 1.5 m, this species can grow rapidly and form dense clumps many metres across in suitable localities. The stems are stout, hollow and visibly grooved. Additionally, the leaf stalks entirely sheath the stem.

In keeping with the plants' size, large umbels containing up to 30 individual flowers are formed. Hemlock Water-dropwort has two levels of umbel structure, giving both primary and secondary umbels. Flowering takes place in June and July.

Hemlock Water-dropwort photographed in mid-June. There are over 20 secondary (terminal) umbels within the overarching primary umbel shown in the photograph. The finely divided leaves of this species are visible on the left of the photograph.

More ... Hemlock, a closely related species to Hemlock Water-dropwort, occurs in the same types of habitats and flowers at the same time. However, Hemlock has stems which are less ridged and typically contain purple blotches.

More ... There are many examples of animals and humans being poisoned accidentally by eating one of these species. There are also examples of deliberate poisoning – Socrates was executed in 399 B.C. by taking a concoction of Hemlock – there is some debate over which of the two species mentioned above was involved.

Some plants cause harm not by being eaten, but by humans making physical contact with them. The rapidly growing and impressively large biennial, **Giant Hogweed**, is such an example. Giant Hogweed is not native to the British Isles but was introduced as an ornamental plant for estates and large gardens around two hundred years ago.

The sap of Giant Hogweed is phototoxic, causing severe blisters and burns if bare skin is exposed to the sap in sunny conditions.

Typically found at river margins, Giant Hogweed is another member of the Carrot family, with large umbels (up to 50 cm diameter) produced at the tip of the plant. Flowers are present during the mid-Summer months and each plant can produce vast numbers of seed ensuring high densities of the plant along river edges unless they are removed.

Giant Hogweed growing at the edge of the River Lagan. A truly massive herbaceous (non-woody) species, plants can reach a height of five metres. The sheer size of the plants makes this a species unlikely to be misidentified.

Alien and invasive species

Aliens may be regarded as species not indigenous (originating) in Ireland. Invasive species are those which spread rapidly, often out of control, through native habitats. Giant Hogweed, discussed in the previous section on poisonous and harmful plants is both an alien and an invasive species. Species such as these did not evolve in tandem with our native species and they have the potential to spread rapidly and cause significant harm to biodiversity.

Himalayan Balsam is a relatively common invasive herbaceous annual. Introduced as an attractive ornamental plant in the Nineteenth Century, it has spread rapidly on river and canal banks and on lake margins. As an annual, it can grow very rapidly in a few months to reach heights of over two metres. The dense thickets it forms enables it to outcompete most native species.

A stand of Himalayan Balsam in mid-July at the edge of Lough Neagh.

The complex flower of Himalayan Balsam. The high rate of nectar production makes this species a popular choice for bees. The top right of the photograph shows some seed pods.

Himalayan Balsam is in flower from July to October. The explosive seed dispersal (when ripe, the seeds will shoot up to two metres if the pods are touched) encourages the rapid dispersal of the species by allowing seeds to reach nearby waterways.

More ... A common name for this species is 'Policeman's Helmet' – if you can imagine the photograph of the flower being rotated, it is not hard to imagine it being the shape of a London 'Bobby's' helmet.

More ... Himalayan Balsam belongs to the small Balsam family (*Balsaminaceae*), mainly consisting of tropical species. The flower structure is very different to that of native NI species – the five petals are highly modified to form the upper hood, lower lip and spur.

More ... Although a non-native alien and having an undesirable effect on native plant communities, Himalayan Balsam is an excellent source of nectar for bees, showing that the conservation approach to non-native species does not always involve straightforward choices.

Perhaps the best known and most feared invasive alien is **Japanese Knotweed**. Japanese Knotweed originates from Asia and was brought to NI in the 1800s. Growing to a height of two metres or more, it can spread rapidly by its underground rhizomes growing laterally extending the area it covers. Japanese Knotweed is in flower from August to October.

Japanese Knotweed photographed in early October.

More … Although the Hemlocks are genuine wetland species, Giant Hogweed and the other aliens discussed above can grow on relatively dry land well away from water bodies. However, they are typically found close to water, largely due to water being a very effective dispersal medium for their seeds. For example, Japanese Knotweed can grow in urban areas, where it can grow through tarmac and the faintest cracks in concrete. This is not surprising, as the species has evolved on volcanic slopes in east Asia where it is adapted to survive being sporadically covered with ash and other debris from volcanic activity.

There are many other alien and invasive species causing harm in non-wetland settings too. **Rhododendron** can spread rapidly through woodland, heath, and bog. Woodland in which **Cherry Laurel** is present in large numbers will have a much-reduced flora. In both cases, the evergreen leaves cast such a dense shade that very few flowering plants can survive on the ground. In addition, each of these species produce toxic chemicals from their roots which create a soil environment unsuitable for many native species.

A final note on wetlands, the central theme of this chapter. The NI climate ensures that there is no shortage of wetland habitats in this part of Ireland. Nonetheless, it is important to note that many of these habitats are fragile and at risk from nutrient enrichment and other pollution. Wetland habitats are also at risk from drainage, reclamation of land for agriculture and urban development.

Wildflowers of Grassland

It's all in the name – grassland is exactly that – a habitat where the dominant species belong to the Grass family. While much of Northern Ireland is covered by agricultural grassland, this is not a natural habitat and is very species poor.

Natural (or semi-natural) grassland is widespread (if not abundant) and usually in small 'parcels' in some parts of NI but surprisingly rare in others. Semi-natural grassland was by far the dominant type of grassland before the arrival of tractors, fertilisers, field drainage and silage making. For natural grassland to be maintained, certain conditions need to be met. These include:

- no reseeding using agricultural grasses – agricultural grasses are more vigorous than native grasses and will outcompete slower growing species

- no fertiliser – fertilisers enable dominance of fast-growing species to the exclusion of the slower growing types

- a soil which is not too rich, for the same reason as above

- some but not too much grazing[1] – grazing will prevent the growth of scrub (shrubs) and eventually trees, but too much will result in significant damage to many species due to being eaten or trampled.

Consequently, many of the best grasslands in NI are found in areas where agriculture is less intensive such as on higher ground, the coastal fringe, thin soils on limestone and in grassland reserves managed by conservation charities or groups.

There are many types of grasslands of course and numerous

---

1 Cutting late in the season after many of the species have set seed or at least have become established such as with traditional methods of producing hay serves the same purpose.

sub-divisions within the major groups. Some of these will be referred to, but grassland types will not be rigidly sub-divided as there is often a large overlap of species composition between different grassland classifications.

Irrespective of type, grasslands tend to be species-rich with many Grass species typically present and that is before other more vibrant flowering species are considered. The next photograph highlights this point.

A species-rich grassland in mid-June; close examination shows that there are several Grass species in this small area together with species from other plant families.

The next photograph shows a section of grassland in a nature reserve not long before being mowed. Note again the species richness.

Most of the yellow flowers belong to Meadow Buttercup. Ribwort Plantain, White and Red Clover and a host of Grass species are present too. Photograph taken in mid-June.

Cuckooflower, also commonly known as Lady's Smock, is a species characteristic of damp meadows and is common throughout NI, although less so in areas with acidic soils. A very pretty, early flowering plant (flowering April-June), it can grow to 50 cm in height. The petals of Cuckooflower can range in colour from white to deep pink or lilac.

A Cuckooflower inflorescence – note the four lilac-tinged white petals and the yellow stamens. The small green sepals are visible in the flowers closer to the tip which are not yet open.

More ... Cuckooflower is an important food plant for the caterpillars (larvae) of the Orange-tip butterfly. The female butterflies lay just one egg on each plant so that caterpillars are not competing for food. The single orange eggs are easier to spot than the green caterpillars!

More ... The name Cuckooflower is derived from its early flowering period coinciding with the arrival of the cuckoo.

Cuckooflower is a member of the large Cabbage family. Species in this family have plants with radially symmetrical flowers each having four sepals and four petals (arranged in a cross) as seen in the photograph. Cross-pollination (by insects) is encouraged by the carpels maturing before the stamens.

Meadow Buttercup is another classic meadow grassland species, preferring damp and non-acidic conditions. This species flowers between April and July and the classic Buttercup-style flowers are pollinated by a range of insect species. Meadow Buttercups are quite a delicate species with flowers at the ends of long stalks, usually held well above the deeply dissected leaves.

Meadow Buttercup with flowers still in bud – note the deeply divided leaves.

The next two species are also classic grassland species, but unlike Meadow Buttercup, they have greater ecological amplitude being able to grow in both calcareous and acidic soils.

Yellow-rattle can be found across NI and each plant can attain a height of up to 50 cm. Distinctive yellow corollas are flattened on each side, with each having a downturned lower lip. Flowering is from May to August and the flowers are grouped in short spikes. As the next photograph shows, its leaves are clearly toothed at edges, are stalkless and organised as opposite pairs on the stems.

Photographed in mid-June, the bright yellow flowers of Yellow-rattle are quite distinctive.

More … Plant names can often tell a lot about a particular species. With Yellow-rattle it is not hard to work out the origin of 'Yellow'. The 'rattle' describes how the seeds rattle in the wind within the seed pods later in the season, giving a distinctive grassland sound. It was said that hay would be ready for cutting when the Yellow-rattle started its rattling.

Eyebrights are more low growing and more sprawling than Yellow-rattle but are also easily recognised. There are many very similar Eyebright species and sub-species which are difficult to separate, but leaving that aside, it is not difficult to recognise an 'Eyebright'. Common Eyebright is, not surprisingly as the name suggests, the most common species.

Flowering from June to September, Eyebrights have white corollas with darker purple veins running across them and a yellow patch on the lower lip. The upper lip of each flower has two lobes and the lower lip three lobes. Flowers are in clusters.

Eyebright photographed in late June. Note the two lobes on the upper lip and the larger three-lobed lower lip (with each lobe clearly indented at its distal edge). The distinctive purple lines and yellow blotches on the lower lip are also clearly shown.

Eyebrights and Yellow-rattle have many features in common, above and beyond their relatively complex insect pollinated flowers and the ability of the species to grow in a wide range of grassland types; they are as likely to be found in dune systems as they are in upland acidic grassland.

They are both parasitic in that their roots penetrate the roots of neighbouring plants and 'steal' some of their nutrients. They are often described as hemi-parasites (partial parasites), as while parasitic activity takes place below ground the above ground parts still photosynthesise normally. Compare this with the (totally) parasitic behaviour of the woodland plant Toothwort. Yellow-rattle and Eyebrights are important grassland plants in that their parasitic activities can stunt the growth of species which might otherwise become dominating thus reducing overall biodiversity.

More … Yellow-rattle seeds are often included in conservation grass mixtures to suppress Grass species which might otherwise become too dominant.

Eyebrights, Yellow-rattle and Toothwort are all classified together in the Broomrape family.

Most of the species covered in this book are perennials, consisting of plants which live for more than two years. Yellow-rattle and the Eyebrights break this trend as they are annuals. Annuals complete their life cycle involving seed germination, growth, flowering and seed production within one year. Annuals are associated with more disturbed (transient) habitats, yet many grasslands do not fit that description. The advantage with Yellow-rattle and Eyebright being annuals is that having been parasitic on (Grass) species in one micro-habitat, their seeds have an opportunity of landing in other sites where there are fresh roots available to parasitise or where there are gaps such as those created by hoof prints. Being annual in essence allows these species to be 'mobile' across the grassland.

Another plant, which while common in a range of grassland types, is versatile enough to be found in more open woodland settings and hedgerows is the patch-forming and sprawling **Germander Speedwell**. A close relative of the rarer Wood Speedwell, Germander Speedwell has darker bright-blue petals and has the two long stamens characteristic of Speedwells in general. Flowering is between April and July. Corollas are about one cm in diameter and are four-lobed with the lowest of the lobes being considerably narrower than the others. As with other Speedwells, Germander Speedwell is a member of the Speedwell family.

Germander Speedwell photographed in mid-June. Note the two long stamens, either side of a much thinner, elongated style, and the four-lobed corolla, with the narrowest lobe at the base.

More ... Germander Speedwell's vernacular names include 'Goodbye' and 'Farewell'. Thought to be a good luck charm for travellers 'speeding' them on their way. In recent centuries, it was also an important herbal medicine for a range of ailments, perhaps 'speeding' recovery.

Another species common in a wide range of grasslands of different soil types is **Common Knapweed** (also called Black Knapweed). A species very common across NI, it can grow up to 70 cm or more.

Flowering is from June to September with composite flowerheads produced singly at the end of stems. Florets range from pink to reddish-purple. Directly underneath the florets, a ring of overlapping dark brown to black bracts with hairy edges helps hold the flowerhead together. With its composite flowerheads, Common Knapweed is a member of the Daisy family.

Easily mistaken at first glance for a Thistle, a close look at the stem will note the absence of sharp prickles or spines which are a feature of Thistles. Nonetheless, the similarity in flowerheads shows that Knapweeds and Thistles are closely related.

Common Knapweed photographed in mid-July. Note the compact flowerhead with many disc-florets packed tightly together. Self-pollinated, Knapweed flowers are an important nectar source for bees and butterflies. Note the hair-fringed dark brown-black bracts immediately below the florets.

The next two species, Selfheal and Bugle are members of the Dead-nettle family, the same family which includes Water Mint and Marsh Woundwort (both covered in the wetlands chapter).

Selfheal is a plant so widespread it is present in the fringes of many household lawns that are not shaved or treated with chemical herbicides.

The inflorescence of Selfheal forms a compact oblong head at the end of stems. Flowers have corollas which are typically violet blue in colour. Each corolla has an upper hooded lip which is more extensive than the lower lip as shown in the next photograph.

Selfheal inflorescence showing individual flowers with a hooded upper lip and a shorter lower lip which is folded back on itself.

Bugle is common across NI in both damp grasslands and damp woods; it has been covered already in the woodland chapter where reference was made to its ability to spread rapidly using runners. Flowers are organised in whorls (rings) creating a terminal inflorescence consisting of a spike of flowers with each whorl forming in a leaf axis. The blue corollas have lower lips with distinct guidelines, but the upper lips are much reduced. Butterflies and moths are important pollinators of Bugle.

Bugle inflorescence (photographed in early June) showing the characteristic bright blue flowers with the lower lip much larger than the reduced upper lip. Note how the whorls of flowers arise at the base (axil) of leaves. The leaves (in opposite pairs) often show a bronze tinge.

The **Common Spotted-orchid** is a common species of damp meadows, other grassland types and even road verges. It is also frequently found in mature dunes but less common in more acidic habitats such as coniferous woods or in soils formed on granite.

Leaves arising from the base of the stem are typically spotted or blotched with (largely) transverse brown-purple markings, although leaves may also be unmarked. Stems reach a height of 50 cm or more.

Flowers are organised in a long spike and open in sequence from the base up. Tepals can be very variable in colour, ranging from white to pink-lilac. The labellum (lower lip) has three lobes, with the central lobe being longest, extending beyond the two lateral lobes. The labellum is decorated with reddish to lilac loops or dashes. Flowering is from May to August. A typical flower spike of the Common Spotted-orchid is shown in the next photograph.

More … It is not hard to work out how this species got its name. It is probably the commonest Orchid – largely due to it being able to grow in a range of different habitat types. The typical intense spotting of leaves has also played a role in its naming.

The Heath Spotted-orchid is so closely related to the Common Spotted-orchid that they frequently hybridise. Also fairly common, the **Heath Spotted-orchid** is at first impression very similar in appearance, having a pretty terminal raceme of white to pink flowers in a cylindrical spike which becomes conical at its distal end.

The narrow lanceolate leaves spreading out from the base of the stem are often splattered with simple spots, less obvious than in its close relative. A key difference between the two species is the structure of the labellum at the base of each flower. In the Heath Spotted-orchid, the two lateral lobes are much wider, making the ratio of width to length of the labellum higher. The middle lobe is smaller and shorter, normally extending no further than the lateral lobes in length and can be almost 'tooth-like' in appearance. The labellum is typically decorated with purple spots and short lines. The flowering period of the Heath Spotted-orchid extends from May to August.

Furthermore, the two species are typically found in different habitats although they can occur together. Whereas, the Common Spotted-orchid is found in meadow grasslands, dune systems and habitats with non-acidic soils, the Heath Spotted-orchid is more common in upland and acidic areas such as moorland, peaty bogs, wet heath and the fringes of coniferous woodland.

OPPOSITE: *Flower spike of the Common Spotted-orchid. Note the white tepals with a hint of lilac and the extensive lilac marking of the labellum. As is characteristic with this species, the central lobe of the labellum extends further than the lateral lobes. This photograph was taken in a species-rich damp meadow in mid-June.*

Heath Spotted-orchid, photographed in early June, in an upland grassland in the Belfast hills. Note the labellum as wide as it is long, with large lateral lobes and a narrow central lobe extending no further than the lateral lobes. Compare this with the shape of the labellum in the previous photograph of the Common Spotted-orchid.

Another Orchid found in grassland is the **Greater Butterfly-orchid**, although it is much less common than the two species just described. This species is characterised by having two large oval leaves at the base of the stem. Stems can be 50 cm or more terminating in a flower spike.

Flowers are white or white with a green tinge. Extending from the base of the tepals is a long and curved spur within which the nectar is often visible at the tips.

As noted earlier in this book, Orchids do not have traditional stamens consisting of distinct filaments and anthers. In contrast, they have a 'mass' of pollen, with pollen accumulations being referred to as pollinia. In the Greater Butterfly-orchid, the pollinia start relatively close together but clearly diverge towards their distal ends.

Flowering is from May to July and pollination is by insects such as butterflies and moths which have a proboscis long enough to extend down into the long spur to reach the nectar.

Greater Butterfly-orchid photographed in mid-June. Note the long strap-like labellum of each flower and the long, curved spurs with nectar visible in their tips. The pollinia can be seen to form a horseshoe pattern as they diverge before straightening near their tips.

Another Butterfly-orchid found in NI is the **Lesser Butterfly-orchid**. Very similar to its close relative, it is usually a bit smaller (more like 30 cm in height) and more suited to heath or bog habitats. However, a diagnostic difference is that whereas the pollinia diverge in the Greater Butterfly-orchid they are parallel in the Lesser Butterfly-orchid.

The **Frog Orchid** is widespread in NI if not common. A species which normally avoids acidic soils, the Frog Orchid typically occurs in the calcareous soils of sand dunes and on limestone and other non-acidic soils further inland.

Frog Orchid photographed in mid-July. At the bottom of the photograph, it is possible to see a stem leaf. The flowers can be seen to have a red-brown tint. The labellum (lip) is very short and indistinct (compared to many other Orchid species).

A species which grows to 30 cm, it can be difficult to spot in vegetation as it blends in with other plants due to its overall greenish colour. Up to five or six broad, blunt, oval leaves form a basal rosette at ground level. Stem leaves arising from the inflorescence stalk are smaller, narrower and more pointed.

Each inflorescence is organised as a raceme and can have up to around 25 flowers, often tinged red or brown, but not to the extent which make them stand out in their habitats. The flowering period is May to August.

More ... It is thought the name Frog Orchid originated because, on close examination, the flowers resemble small frogs in their pre-jumping pose.

The **Common Twayblade** is an Orchid which can be difficult to spot in grassland due to its green colour. Usually somewhere between 20-70 cm in height, plants of this species have two large ovate leaves near the base of the stem.

Each plant has its yellow-green flowers arranged loosely on a long spike with the flowering period extending from June to August. Five of the six tepals form an upper hood which contains the pollinia and the sixth tepal is extended into a strap-like labellum (lip) which stretches well below the rest of the flower and is deeply divided into two lobes at its distal end.

Several plants are often found together due to lateral spread by short rhizomes. Common Twayblade is widespread due to its ability to grow in a range of different habitat types, including woodland and dune slacks, as well as grassland. Unlike most other Orchids (or most other flowering plant species) it is comfortable in each of full sunlight and heavy shade and can grow in both acidic and alkaline soils. The next photograph shows the flowering spike of Common Twayblade growing in a grassland.

Common Twayblade photographed in mid-June. Note the widely spaced flowers (compared to most Orchids). The deeply divided labellum (lip) can be seen extending down from the base of each flower.

Dune slacks and grassland areas between the more mature dunes are in effect mini grasslands. In addition to the species described in the section on dune systems covered in an earlier chapter, other interesting species occurring on calcareous grasslands include the **Bulbous Buttercup**, a species found in dry fast-draining grassland.

As with other Buttercups, the Bulbous Buttercup has five yellow petals and an open simple flower with flowering taking place between April and June. Due to its more specific habitat requirements, it is much less common than the Meadow Buttercup or another well-known Buttercup, the Creeping Buttercup, which will be covered later.

Growing to around 40 cm, the leaves of the Bulbous Buttercup are very like those of the Meadow Buttercup, so when not in flower it can be difficult to distinguish between the two species. However, when in flower, the giveaway is that in Meadow Buttercup, the yellow-green sepals are positioned tightly against the underside of the petals, whereas in the Bulbous Buttercup they are clearly reflexed back on themselves (downturned) as shown in the next photograph.

A Bulbous Buttercup flower showing the classic Buttercup corolla with five yellow petals. Note the downturned yellow-green sepals, a key feature when identifying this species.

Another species more suited to well drained calcareous soils such as those found on limestone in west Fermanagh (rather than the soils covering much of the east of NI) is the Cowslip. **Cowslips** are closely related to the Primrose and as such are in the Primrose family. As with Primroses, they have a rosette of crinkly leaves at ground level.

Flowers are orange-yellow, and small numbers are grouped together in a cluster at the top of an upright stalk which extends from the centre of the rosette of leaves. Flowering is during April and May. The base of the corolla forms a tube which opens at its distal end. Cowslip flowers have a characteristic drooping appearance.

As with the Primrose, cross-pollination in Cowslip is encouraged by flowers exhibiting **heterostyly**. In each of these species, the flowers of around half the plants have a style which is longer than the stamens, meaning that the stigma is at the opening of the corolla tube (with the anthers remaining deep within the tube). In the flowers of the remaining plants, it is the anthers which are projected forward to the edge of the corolla tube (and the stigma remains deep in the corolla due to the shorter length of the style). Flowers which have the stigma projected are described as being of the 'pin' form, with those having the stamens projected being the 'thrum' form. The consequence of this is that 'pin' flowers are effectively female and receive rather than donate pollen when visited by insects and the 'thrum' flowers are effectively pollen donors (males).

The two forms of Cowslip flower discussed above are shown in the next two photographs.

Cluster of Cowslip flowers showing the 'pin' form with the stigma at the opening of the corolla tube.

Cluster of Cowslip flowers showing the 'thrum' form with the stamens being projected to the opening of the corolla tube.

Most people can recognise a Dandelion, but few realise there are very many 'Dandelion-like' species with yellow flowers and leaves organised in a low growing rosette. Many of these species are very difficult to identify, but an exception is Mouse-ear-hawkweed, a species common across much of NI.

Mouse-ear-hawkweed is found on well drained soils such as coastal sandy calcareous soils, dry grassy banks, and rocky outcrops and ledges. With composite flowers, it is a member of the Daisy family, and has a flowering period running from May to October. Grey-green, very hairy leaves are arranged in a ground-hugging basal rosette. A flower stalk with one terminal flowerhead (3-4 cm in diameter) can extend to 30-40 cm above the leaves. Flowerheads are formed of lemon-yellow ray florets, with some of the outer rays being red on their undersides. The next photograph, taken in mid-June, shows Mouse-ear-hawkweed growing in rapidly draining short turf in an area of grassland above a coastal cliff.

Note the very hairy grey-green leaves of Mouse-ear-hawkweed in a low rosette and the lemon-yellow flowerhead with some of the outer rays having a red underside tinge. There are leaf sections and sections of flowerhead stalks of other plants of this species in the photograph, probably initially connected by creeping runners. This is an ideal habitat for Mouse-ear-hawkweed as its low growing leaves are not shaded by other plants.

More ... Rosettes have advantages, not least the very low growing habit which protects to some degree against grazing animals. However, plants which have leaves in rosettes only, are susceptible to being outcompeted by those species which grow higher and place them in permanent shade. For this reason, species with rosettes are more likely to be found in short-turf grasslands or dune slacks rather than many other habitats such as woodland.

The next photograph is also of Mouse-ear-hawkweed and this image emphasises how the upper part of the flowerhead stalk is covered by short dark hairs as is the underside of the involucre (the rosette of dark bracts supporting the florets).

Mouse-ear-hawkweed showing the mass of hairs on the involucre and the stem below it.

More ... Hairs have many roles in plants. The primary function in many species is to reduce air flow over leaves and thus reduce water loss by evaporation. However, in Mouse-ear-hawkweed the distribution of hairs in the upper part of the flowerhead stalk suggests a different function. Flowerheads rich in nectar and nutritious pollen will be an attractive food to many animals but having to ingest lots of hair may act as a deterrent.

Devil's-bit Scabious was touched on in an earlier chapter. A common species in a wide range of grasslands, it has multiple small blue to violet-purple flowers tightly packed into dense flowerheads. Magenta stamens are prominent and extend well beyond the four-lobed corolla tubes, with each of the four lobes being approximately of equal size. Flowering is from June to October and pollination is by bees and butterflies.

Note the magenta stamens extending beyond the equally sized, tightly packed flowers in a pin-cushion type flowerhead of Devil's-bit Scabious.

More ... The flowers of Devil's-bit Scabious are nectar rich and provide nutrition for a wide range of insects. It is the main food plant for the caterpillars of the Marsh Fritillary butterfly.

Devil's-bit Scabious is a relatively late flowering species common across NI, so when in flower can be very striking in grassland, whether in damp rush pasture or drier calcareous coastal grasslands.

Devil's-bit Scabious catches the eye on this clifftop in late August.

Field Scabious is closely related to Devil's-bit Scabious but is less widely distributed, preferring neutral or calcareous soils. It is most likely to be found in coastal areas, being relatively rare in the more upland Mourne or Sperrin Mountains or on the Antrim Plateau.

Although similar in many ways to its close relative including flowering period, flowerheads being raised well above the basal leaves, the same flower colour, and the presence of prominent stamens, there are subtle and more obvious differences. Differences between the species include the flowerhead being more flattened and slightly wider and the leaves being divided rather than entire in Field Scabious.

Field Scabious flowerhead photographed in late August, showing the flatter flowerhead compared to its close relative.

Another widespread, late-flowering species found in grasslands, and in a range of other habitats including waste ground, road verges and sand dunes is **Common Ragwort**. A species with irregular and jagged-edged leaves, it typically grows to a height of between 0.5–1 m. A member of the Daisy family, Common Ragwort has composite flowers with the disc-florets and ray-florets both yellow in colour. Flowers are in dense clusters of flat-topped flowerheads.

Common Ragwort is the species of choice for Cinnabar moth caterpillars and plants are commonly infested with the yellow and black caterpillars of this species, which feed voraciously on the leaves.

Cinnabar moth caterpillars on the flowerheads of Common Ragwort.

Due to the presence of toxic alkaloids, Common Ragwort is poisonous and if eaten in high concentrations can be fatal to horses and cattle. For this reason, farmers have been encouraged to prevent a build-up of this species on their land.

More ... As a result of feeding on Ragwort, the caterpillars too are poisonous. They advertise this fact by their black and orange colouration which warns off potential predators.

More ... The number of caterpillars on a Ragwort plant can reach high enough levels to lead to all the leaves on the plant being eaten, a situation which is hardly ideal for either the caterpillars or the plants! Compare this with how the Orange-tip butterfly lays its eggs on Cuckooflower as noted at the start of this chapter.

Many of us of a certain vintage, and having experienced a rural upbringing, remember a childhood of play in meadows and their associated smells, sights, and sounds. Meadows are much less common now because of the move towards intensive farming in recent decades, with species-rich meadow grasslands becoming a genuine rarity, certainly in the east of the country. Nonetheless, a trip through grassland, coupled with a basic understanding of the species we are passing through can still bring rich rewards.

Wildflowers of Heath, Bog and Moorland

When we think of heath, bog and moorland we tend to think of bleak, windswept and wet upland areas remote from centres of civilisation. In many ways this is true, as they are habitats not normally suitable for habitation, and often well away from the main urban centres.

Heath forms a habitat in which shrubs such as Heathers form the dominant vegetation. Soils are typically of poor quality and acidic. Later in this chapter, we will distinguish between dry heath (where the soil is well drained) and wet heath (where it is not).

Bog is generally much wetter than heath and is formed on waterlogged, acidic peat. There is some overlap with heath in that Heathers can be common, but a key feature is that Sphagnum moss is ubiquitous in bog. Bog is often referred to as peatland.

There is often a gradation between these habitats with bog often appearing in the wetter areas within heathland. More on moorland later.

Heath

Heaths typically have few or no large tree species present with Heathers normally being the dominant flowering plants. Heathers are woody, evergreen shrubs belonging to the Heather (*Ericaceae*) family.

Heather or **Ling** is probably the most common Heather species in Northern Ireland. Although typical of both dry and wet heath, it is also found in some woods and bogs, and potentially anywhere the terrain is suitably infertile and acidic.

Heather has small overlapping flat leaves arranged in dense rows, and the small flowers range in colour from pink to purple. The flowers are organised in dense spikes with flowering taking place from July to September or October.

A flower spike of Heather. In this species the sepals are longer than the petals, so the colour is provided by sepals rather than petals. Note the short, tightly packed overlapping leaves.

Heather photographed in mid-August, on rapidly drying soil on limestone. Note the numerous flower spikes extending out from the body of the plant in all directions.

More ... Heather (often with other twigs) was used to make brushes before they were commercially produced. Historically, they were known as besoms.

Bell Heather is common in dry heath, but much less so in wet heath, being more restricted than Ling in habitat preference. Growing to around 50 cm, Bell Heather has short (around 5 mm), very thin leaves in whorls of three along the stem. Flowers are a magenta-purple shade, and are positioned at the end of stems with the corolla being noticeably longer than the short sepals, in contrast with Ling. The 'bell' is formed of petals fused together. The organisation of the short leaves (in whorls rather than long flattened rows) and the reduced number of larger bell-shaped flowers positioned terminally, all make Bell Heather relatively easy to tell apart from Ling.

Bell Heather photographed at the end of August. It is in flower from June or July to September. Note the magenta-purple, bell-shaped flowers positioned at the end of the stem. When mature the style extends through the rim of the corolla 'bell'.

The third species of Heather commonly found in NI is **Cross-leaved Heath**. Characteristic of wet rather than dry heath, Cross-leaved Heath is also a classic bog species where infertile, very wet acidic soils are also found. It is closer in appearance to Bell Heather than it is to (Ling) Heather, being of a similar size, with similar shaped 'bell' flowers and thin spiky leaves organised in whorls at intervals along the stem.

However, in Cross-leaved Heath individual leaves are in whorls of four (rather than in three groups as in Bell Heather), and although the flowers are also in terminal clusters, they are a paler shade of pink.

Cross-leaved Heath photographed in early March. Note the grey-green leaves organised as whorls with four thin spiky leaves per whorl clearly angled upwards. The flowers are from the previous summer (flowering is from July to September), but their bell shape and position at the tip of stems is evident.

More … The scientific name for Cross-leaved Heath is *Erica tetralix*. In Greek, *tetralix* means leaves in fours.

Gorse (covered in detail elsewhere in this book) and the rapidly spreading fern Bracken are frequently present in dry heath habitats. However, in some NI heaths, it is Western Gorse rather than the overall more common (European) Gorse which is present.

A dry heath in late September. Gorse is still in flower, but the Bracken (left and top of photograph) and the Bell Heather (bottom right) are well past their best.

While heath is dominated by a shrub layer, there are several species of flowering plants typical of this habitat. **Tormentil** is a very common species of both dry and wet heath, although it is also common in acid grassland, hedgerow and on some calcareous soils, showing that it has what can be described as extensive ecological amplitude. A member of the large Rose family, it is slightly unusual when compared to many of the other roses in that it has four rather than five petals, each with an almost flattened distal edge that is typically faintly indented.

Tormentil's bright yellow flowers are around one cm in diameter and the plant is in flower between May and September.

Tormentil photographed in mid-August. Note the flattened outer edge of the petals and the gaps between adjacent petals making the shorter sharp-ended sepals visible. The leaves are deeply divided as shown.

Heath Bedstraw has the small four-petalled flowers (approximately 3-5 mm in diameter), and leaves arranged in whorls at intervals along the stem typical of plants in the

Bedstraw family. Heath Bedstraw has a prostrate mat-forming habit, seldom extending more than 20 cm above the ground.

Normally in flower from June to August, Heath Bedstraw is very common in heath habitats, particularly in drier sections and is also common in acid grasslands.

Heath Bedstraw photographed in late June. Note the mass of small flowers with four white petals grouped in dense clusters.

Another species found in dry heath and present in many grasslands is the very pretty **Heath Speedwell**. While the small leaves with serrated edges crawl over the ground surface, the most visible feature is the vertical spike of flowers which extends to around 10 cm in height.

A species very easily overlooked, the corollas are pale lilac and usually between 5-10 mm. As with all Speedwells, each flower has two stamens and a style which extend beyond the corolla. Flowering is from May to August.

Heath Speedwell photographed in mid-June. The vertical spike is a raceme with the small lilac flowers opening in order up the spike. This photograph was taken in a dry heath at the point where the heath was merging into habitat more typical of grassland. The unopened red-yellow flowers on the right are Common Bird's-foot-trefoil, a classic dry grassland or dune species.

Two contrasting examples of Northern Ireland heath habitats
The slopes of the **Eastern Mournes** provide an excellent example of heath habitat. Being on mountainside, the full range from lowland dry heath to upland wet montane (mountain) heath is present. Soils are typically shallow and peaty on the granite slopes.

The Heather family species covered earlier in this section are present. Other interesting species include **Western Gorse**, a species which is typically half the size (up to around 50 cm in height) of the more common Gorse and is much more compact than its typically sprawling close relative. Flowering is normally from July to October – a shorter flowering season than Gorse, which is typically at its best in Spring. Western Gorse is not restricted to the Mournes, being found on the slopes of Slieve Croob in mid-Down and in the Sperrin Mountains. Other species include the relatively rare shrub Juniper and the much more common Bilberry, a close relative of the Heathers.

More ... Bilberry is known in many localities as Blaeberry. Their berries are a popular ingredient in dessert pies.

Geographically close to the Eastern Mournes heathland, parts of the **Murlough sand dune system** provide a very different example of dry heath habitat. The younger dunes closer to the sea have a classic dune flora on calcareous soils. Further inland the dunes are much older and much more stable. The once-calcareous soils in this section have been leached by thousands of years of rainfall to the extent that the shallow soils are deficient in lime and essential minerals, creating what is described as a base-poor soil which is much more acidic than the soils closer to the sea.

Gorse and Bracken are very common on these dry soils, as is Heather and Bell Heather.

More ... Very unusual for a heath habitat, Marram grass grows amongst the Heathers, Gorse and Bracken at Murlough. The Marram colonises gaps which appear because of erosion.

An interesting plant of dry heathland seen at Murlough is **Wood Sage**. Wood Sage has wrinkled, oval leaves, not unlike its relative, herbal Sage. Growing to around 50 cm, and flowering between July and September, it is a good source of late-Summer and Autumn nectar.

Wood Sage, photographed in late September (just outside the flowering period) in dry heath. Note the crinkly, rough leaves. Leaves are in opposite pairs as is typical of the Dead-nettle family. Wood Sage is a good indicator of acidic soils.

Wet heath can be difficult to distinguish from bog as both are found on peaty soils. Wet heath is typically in more upland sites than dry heath and has a different range of species. Cross-leaved Heath is much more common and Bog-myrtle is also found. Some of the species covered in the next section on bog also occur in wet heath.

In summary, heath is characterised by poor quality, usually acidic, soils and a dominating shrub layer, typically containing various Heathers and/or Gorse. Grazing, cutting, or burning is usually necessary to prevent the growth of trees and succession to woodland. Burning can be spontaneous due to the build-up of wood in habitats which can become very dry in Summer, or due to carelessness, reckless or deliberate behaviour.

Other classic heath species include Heath-spotted Orchid (covered in the grassland chapter), Lousewort and Devil's-bit Scabious.

More … Heather and Gorse species have narrow leaves. This is an adaptation which reduces water loss by reducing the area across which water can evaporate out of leaves. Although dry heaths can experience relatively high rainfall, the shallow soils are very poor at retaining moisture and therefore dry quickly.

Bog

A bog is formed on relatively flat, poorly drained ground in which there is a build-up of peat as the substratum in which plants grow, rather than in mineral soils. Peat 'soils' are acidic, nutrient-poor and waterlogged or at least very wet. The peat is formed from accumulations of Sphagnum mosses, which when dead, build up as decomposition fails to take place in the very acidic, waterlogged environment. Although wet heath can have peaty soils, in bog the peat is typically much thicker.

Large **blanket bogs** are present in many areas of NI, including parts of the Antrim Plateau (particularly the Garron area), in the Sperrin Mountains, and on Cuilcagh Mountain in Fermanagh. Blanket bog is most extensive on higher land (above 200 m).

Lowland raised bog develops primarily on lower land than blanket bog. Lowland raised bog can form on river flood plains and on low-lying areas where drainage is very poor. They are typically less extensive in area than blanket bogs, but some raised bogs such as the area southwest of Lough Neagh known as Peatlands Park cover a vast area.

Patches of vegetation alternate with pools in this lowland raised bog. The greener vegetation is Sphagnum moss, and the brown-green tussocks are formed of Purple Moor-grass. Purple Moor-grass is typically tussock forming and it is abundant in a range of wet habitats including bog, wet heath and even wet grassland.

Localised bog habitats also occur in mountain areas where wet heath merges into bog as drainage becomes poorer and the ground more waterlogged. Examples of this can be found in the western reaches of the Mourne Mountains.

A classic bog species is **Bog Asphodel**, a species never found on calcareous soils. Bog Asphodel also occurs in wet heath. It can grow in large stands, often through a combination of rhizome spread and seed. A very pretty plant, which can grow to a height of 40 cm, it has sabre-shaped, flat leaves, curved inwards at their tips originating from ground level.

Inflorescences of Bog Asphodel photographed in early July. This species flowers in July and August.

The yellow star-shaped flowers are spread along a terminal raceme. The orange-red anthers extend from hairy (woolly) filaments. The next photograph shows an inflorescence spike in greater detail. In Autumn, during fruit formation, the entire plant becomes a deep orange colour which is every bit as striking as the plant in flower.

More ... The Latin name for Bog Asphodel is *Narthecium ossifragum*. 'Ossifragum' means bone breaker and it is thought this name originated when it was assumed that sheep feeding extensively on Bog Asphodel developed brittle bones. In reality, it was the lack of calcium in the acidic peaty soils, resulting in a reduced calcium diet which led to the brittle bones rather than feeding on a particular plant species.

Note the star-shaped flowers of Bog Asphodel, each with six narrow tepals. The 'woolly' filament and orange anther of each stamen are quite distinctive.

More ... Bog Asphodel is the sole species in the *Bog Asphodel* family, making it a family of one. Compare this to the numbers in some of the other families covered so far.

Cottongrass, perhaps better known as **Bog Cotton**, is an overarching name given to a small number of species in the Sedge family (*Cyperaceae*), species associated with bog and wet heath habitats. Common Cottongrass and Hare's-tail Cottongrass are the species found in NI.

These plants have slender stems with a white cottony head of flowers. The 'cotton' heads are perianths consisting of bristles rather than typical sepals or petals. **Common Cottongrass** has terminal inflorescences of 3-7 heads, with the perianth bristles being up to five cm in length. **Hare's-tail Cottongrass** has a single terminal white tuft.

Common Cottongrass photographed in mid-July. Note the presence of four separate spikelets (tufts) on the stem, thus identifying this species as Common Cottongrass (rather than Hare's-tail Cottongrass).

Carnivorous or insectivorous species are very unusual plants and are particularly associated with bog habitats. The most common insectivorous species in bog habitats are the Sundews, with Round-leaved Sundew being by far the most common. **Round-leaved Sundew** is widely distributed in suitable habitats across NI, ideally suited to wet, acidic bog, typically growing in association with Sphagnum moss.

Round-leaved Sundew is one of those species where the leaves are more interesting than the flowers. Each of the round leaves at the end of long stalks is approximately one cm in diameter when fully grown. The leaves have a covering of sticky, red-tipped, gland-bearing hairs. Insects landing on the leaves are trapped by the sticky substance produced by the glands. The leaves then curl around the insect, further trapping it and making sure it has no chance of escape. Enzymes released by the leaves digest the insect, with the insect's nutrients compensating for the poor nutrient status of the bog.

A small number of small white flowers are sited terminally on an erect inflorescence stalk which extends up to 15-20 cm above the basal rosette of leaves. Each flower-bearing stalk arises from the centre of the rosette and the flowering period runs from June to August.

Round-leaved Sundew growing in a bog in the Mourne Mountains – photographed early-July. The red-tinged circular leaves on long stalks arise from a central point. Note the inflorescence stalk arising from the centre of the rosette. At its tip, there are several unopened flowers.

More ... The sticky ends of the leaf hairs look like dew drops. Unlike, morning dew, these drops of liquid do not evaporate as the day warms up and are still present in full Sun, hence 'Sundew'.

Two other species of Sundew are found in NI bog habitats, each having the same flowering period as Round-leaved Sundew, although each is much less common than Round-leaved Sundew. The insectivorous way of life is similar in the other two species.

Great Sundew has leaves up to 10 cm or more in length (including the stalk). The leaves narrow gradually into their stalks, which are usually much longer than the leaf blades. In Great Sundew, as with Round-leaved Sundew, the inflorescence stalks arise from the centre of the rosette.

Great Sundew growing in a bog south of Lough Neagh, photographed in early July. The overall size of the plant and the leaf shape, indicate that this is Great Sundew rather than Round-leaved Sundew.

Oblong-leaved Sundew, the third Sundew species in NI, has spoon-shaped leaves which narrow more sharply than in Great Sundew into their stalks. In Oblong-leaved Sundew, unlike the other two species, the inflorescence stalk arises from below the rosette and curves out and upwards before extending above the rest of the plant. When fully grown, Oblong-leaved Sundew is more similar in size to Round-leaved Sundew than the larger Great Sundew. All three species belong to the small *Droseraceae* (Sundew) family.

Butterworts are another group of insectivorous plants, albeit very different in appearance to the Sundews. As in the Sundews, Common **Butterwort** has a basal rosette of leaves but in this species, there are no leaf stalks, and the elongated ovate-oblong leaves are very different in shape and colour. The leaves of Common Butterwort are yellow-green and are up to 7-8 cm in length when fully grown. The upper surface of the leaves has sticky glands which trap insects that land on it. Once an insect is trapped, the leaf edges curl in slightly resulting in increased surface area of leaf in contact with the doomed prey. As in Sundews, enzymes released by the leaf digest trapped insects with their nutrients subsequently being absorbed by the plant.

Common Butterwort with characteristic yellow-green leaves.

Single flowers are produced at the top of each long flower stalk which may reach 20 cm in length. Common Butterwort is in flower between May and July and the violet flowers are pollinated by bees. Common Butterwort is a member of the Bladderwort (*Lentibulariaceae*) family.

More … The scientific name for the Common Butterwort is *Pinguicula vulgaris*. *Pinguicula* means 'buttery or sticky leaf' in Latin.

Other bog species typical of bog habitats include the shrubs Bog-rosemary and Bog-myrtle.

Bog-rosemary, a member of the Heather family is a species of lowland raised bog. It is rare and only found in a few NI sites. Bog-rosemary is a low growing shrub (seldom more than 30 cm in height) which produces pale pink flowers that hang in small clusters on long stalks from May to June or July. Bog-myrtle is also found on heath as described earlier.

Bogbean, a herbaceous perennial, is in flower from March through to June and is found in bog pools and other wetter parts of bog. Bogbean encourages cross-pollination by its plants showing heterostyly (as do Cowslip and Primrose). A relatively common species, Bogbean is not restricted to bog habitats, being found on lake edges and in marsh and other suitable wet habitats.

In some areas of bog or even in wet heath, where the land is sloping or where a spring reaches the surface a **flush** habitat is present. Flushes are habitats where there is lateral movement of water over the ground, rather than the water being stagnant (but not to the extent of forming a stream or river). Some species are particularly associated with flushes, including Common Butterwort and the rare Bog Orchid.

Sphagnum moss has been referred to several times in this chapter. Sphagnum is the plant most associated with bog (and even wet heath), and it is Sphagnum which has not properly decomposed that is the main component of peat. In 'living' bogs, it is mainly the layer of Sphagnum moss at the surface which forms the 'living' component. The next photograph shows both Great Sundew and Round-leaved Sundew growing on a rich carpet of Sphagnum in a bog, emphasising the significance of Sphagnum mosses in these habitats.

BELOW: *Sundews on a bed of Sphagnum moss, photographed in early July. Centre and left are two Great Sundew plants with a much smaller Round-leaved Sundew in the right background. The Great Sundew in the centre of the photograph appears to have trapped insects in at least two of its leaves. In each plant it is possible to see inflorescence stalks unfurling.*

More … Sphagnum is very effective at soaking up and retaining water. Many gardeners use it as a lining for hanging baskets and in tubs to avoid the containers drying out too quickly.

More … Sphagnum has antiseptic properties and in World War One it was used by medical staff to help reduce infection in wounds. Sphagnum was collected from the Peatlands Park area, the Mournes, and other sites for this purpose.

Bogs can be very diverse habitats, often due to human activity. The large-scale removal of peat for fuel or compost, produces sections of cut-away bog which create a special range of habitats of their own, quite different to untouched 'living' bog. Wetter and drier parts of bog have a very different range of species too.

Damaged bog, due to human activity, is prone to drying out and at risk of colonisation by species such as Rhododendron and **Rosebay Willowherb** – species which are excellent colonisers of a range of habitat types.

Rosebay Willowherb photographed in mid-July. A species which can colonise large areas in a short period of time due to rhizome growth in addition to seed spread. As seen in the photograph, flowers are in terminal racemes. Flowering is from July to September with pollination by bees. Closely related to Great Willowherb (covered in the wetlands chapter), these two species belong to the Willowherb (Onagraceae) family.

More ... Peatlands Park contains excellent examples of 'living' bog and sections of cut-away bog, which are currently undergoing restoration. Many of the classic bog species are found within this site. A big advantage of studying bog species in Peatlands Park is that they can be seen from the sanctuary of dry paths!

More ... Peatlands Park is not unique in opening access to bog habitat. The Cuilcagh Boardwalk Trail in County Fermanagh traverses one of the largest areas of blanket bog in NI and plans for making bog habitat in other areas more accessible to the public are in the pipeline.

Large masses of peat, such as in the extensive blanket bogs, act as valuable carbon stores where carbon is 'locked up' within the peat. The carbon will remain locked up if the peat is prevented from decomposing.

More ... Poor management of bog, such as the large-scale removal of peat for use as commercial compost and as an alternative to wood or coal in domestic fires, results in the drying out of parts of the bog leading to decomposition, thereby releasing carbon dioxide which contributes to global warming.

More ... There is growing evidence than many European bog habitats are beginning to dry due to climate change, irrespective of human activity.

Although included in the chapter title, there has been little reference to Moorland in this chapter. **Moorland** is an overarching term covering upland wet heath, bog on higher ground and even upland acidic grassland. Conditions are invariably wet and cold. Depending on the actual habitat involved, the species present change, but species such as Cross-leaved Heath are common.

Having a wet and cool climate, together with the fact that we have many upland areas with peat or poor-quality soils, it is hardly surprising that this part of the world is well endowed with heath, bog and moorland.

A Year in the Life of a Hedgerow

Except when on relatively high ground such as in the Mourne or Sperrin Mountains, or in parts of the Antrim Plateau, we are never too far from a hedgerow. Hedgerows come in all shapes and sizes, but they invariably have a boundary or barrier function.

For the purposes of this chapter, the hedgerow will refer to those bordering roads (rather than those separating fields) and will include the woody species which form the barrier. The bank or verge between the barrier and the road is also included.

January and February

In the early part of the year, hedgerows have relatively few plants in flower. It is a time when the hedgerows themselves are often harshly cut back using tractor-based cutters. From a conservation perspective, this reduces the chance of the hedgerow growing to the extent that more complex habitats develop, but it does increase the amount of light reaching the ground at the base of the hedge, which does benefit some hedgerow species.

Apart from the degree of trimming affecting species richness, hedgerows backing on to woodland are likely to show more biodiversity than those bordering agricultural fields. The hedgerows shown in the next photograph are typical in appearance to those at the start of a calendar year, showing hedges dominated by Hawthorn and populated by mature trees at intervals. Although Hawthorn tends to be dominant in many hedges, many other species are typically present in the stock-proof barrier including Blackthorn, Hazel, Gorse, Holly, and a range of Willow species.

More ... Most hedgerows have their origins as planted hedges for barrier or boundary purposes and are consequently species poor, as only those species which form effective barriers were used. However, some are remnants of ancient woodland habitats and are much more species rich.

Country hedgerows at the start of the calendar year.

There can of course be welcome surprises. A short row of mature Beech is a welcome sight in this hedgerow.

A row of mature Beech in this hedgerow.

In the early months, Ivy and a range of Grasses are the most obvious flowering plants, although neither are in flower then. However, there may be a considerable number of berries still on Ivy at this stage. Most herbaceous flowering plants are still overwintering as underground storage (perennating) organs such as bulbs, root tubers and rhizomes in these months.

Nonetheless, some Spring species including Lords-and-Ladies and Wild Garlic are beginning to push their leaves through the soil and leaf litter. The next photograph shows how **Lords-and-Ladies** has its leaves tightly folded into 'spear-like' structures as they penetrate the soil. Wild Garlic leaves are also tightly coiled as they emerge from the soil surface; the underlying strategy being that the upward force is concentrated into as small a point as possible, thus both maximising force through the soil, and minimising the risk of damage to delicate leaves.

Once the leaves are safely above ground the leaves rapidly expand to make use of the abundant light in these early months before the hedgerow trees and shrubs come into leaf.

Lords-and-Ladies leaves are highly adapted for penetrating the soil and leaf litter. Photograph taken in February.

By February, the leaves of Cow Parsley, Lesser Celandine, and Creeping Buttercup are all beginning to appear.

Spring brings a lot of change, with most hedgerows bursting into life from March on. The next section highlights some of the changes which take place over this period.

March and April

During these months, there is significant activity in the barrier species themselves. Several hedgerow species including **Hazel** produce catkins as early as February or March. The next photograph shows the male catkins of Hazel, a common hedgerow tree.

Photographed in mid-March, the creamy male catkins of Hazel hang well below their supporting branches. Female flowers are much smaller and mature slightly later than the male flowers thus increasing the chances of cross-pollination.

By the middle of March, **Blackthorn** is coming into flower and like many species in the hedge barrier it is a dense spiny shrub which provides considerable resistance to penetration. Blackthorn is a member of the Rose family.

The white flowers of Blackthorn add early colour to hedgerows from March.

Many people giving the white blossom of Blackthorn a cursory glance in March or April assume that the species responsible is Hawthorn. However, with care it is relatively easy to tell these two species apart. While Blackthorn will come into flower in March (or early April at the latest), Hawthorn is unlikely to produce its blossom until mid-May. Furthermore, when Blackthorn comes into flower its leaves are not fully expanded. Additionally, the leaves of Blackthorn are more ovoid in shape and have a slightly toothed edge. By the time Hawthorn breaks into flower, its leaves are already fully expanded, and the deeply lobed leaves are very different in shape to those of Blackthorn.

Blackthorn flowers have five petals and numerous stamens with brown anthers at the end of long filaments. Note how the leaves have not yet opened, but the presence of gently toothed margins at their edges can still be seen (top right). Photograph taken in early April.

By mid-March, the ground cover has increased dramatically. Lords-and-Ladies leaves have expanded to full size and many more Lesser Celandine plants are in flower. Some Primroses are in flower by then too.

Common Dog-violet occurs throughout NI and is by far our commonest Violet. Producing small blue flowers from March-May, this species occurs on hedgerow banks, although it is also present in woods and grasslands. As with other members of the Violet family, the five-petalled flowers are bilaterally symmetrical and solitary rather than being in an inflorescence. The lowest petal extends back into a spur; in this species the spur curves up and is paler than the petals and has an obvious notch (indent) running along its length.

Common Dog-violet photographed in mid-April. Note the upward pointing, cream coloured spur (the central notch on the spur is not visible as it sits directly under the downwardly curving flower stalk in this photograph). Note the similarity in petal arrangement with the closely related Wild Pansy.

More ... The addition of 'dog' to a plant's name indicates that it was regarded as an 'inferior' species. Although unscented, no one could deny that Common Dog-violet is very pretty.

Garlic Mustard (also called Jack-by-the-hedge) is a biennial common in some areas and is in flower from as early as April. A species which can grow to heights of one metre, it typically forms patches of plants close to the barrier species in a hedgerow, presumably as they provide a degree of protection against wind damage that is not present further out on the verge. A member of the Cabbage family, Garlic Mustard has its classic floral arrangement of four white petals arranged in a cross. The flowers are positioned in terminal clusters.

Garlic Mustard photographed at the end of June. Note the heart shaped, crinkly leaves with mildly serrated edges. The small white flowers are in terminal clusters.

More … The reference to 'garlic' in this plant's name is because the leaves when crushed smell of garlic. Although Garlic Mustard has this in common with Wild Garlic and commercial Garlic, it is not closely related to the more well-known species.

Ground-ivy (a very different species to Ivy the climber, and a member of the Dead-nettle family), Wild Strawberry and Barren Strawberry (both members of the Rose family), all come into flower during this time. Wild Strawberry and Barren Strawberry each have small five-petalled white flowers, but the obvious gap between the petals in Barren Strawberry means these two species are easy to tell apart. Herb-Robert is also a common species in hedgerows from Spring through the Summer and into Autumn.

May and June

Hedgerows are perhaps at their prettiest during May and June. There is no doubt they are most interesting at this stage with both the hedges and the verges brimming with colour.

Hawthorn provides an excellent barrier due to its thick growth and the presence of thorns. It has deeply lobed leaves, which clearly distinguishes it from Blackthorn, white (occasionally pink) flowers with five petals and numerous stamens. Flowering in Hawthorn is during May and June.

Hawthorn in mid-May – the photograph shows that many of the flowers are not yet open. As with Blackthorn, there are five white petals and numerous stamens in each flower. Note the deeply indented leaves.

Gorse, a large evergreen shrub when fully grown, is often present as part of the hedgerow barrier due to its sharp spines being effective in retaining stock. Being in flower for much of the year, although typically most vibrant in Spring, Gorse adds colour even in the winter months.

Gorse is a particularly common hedge species in parts of the country where the land is rocky and soils are poor, such as in more upland areas, as it is well suited to growing in these conditions.

Note the sharp spines on this photograph of part of a Gorse shrub.

More ... Gorse, in common with other members of the Pea family, can capture nitrogen from the atmosphere through nitrogen fixing bacteria found in its root nodules. The nitrogen is then used to make protein in the plant. This ability to use atmospheric nitrogen is crucial in its ability to survive in poor quality, nitrogen-deficient soils.

Broom is rarer than Gorse and is a shrub of similar size occasionally found in hedges. However, the absence of spines makes it a much less effective barrier. Broom has a much shorter flowering season than Gorse, normally being in flower from April to June.

*Broom photographed in mid-May. Note the absence of spines.
Note how similar the flowers are to those of Gorse. Unsurprisingly,
Broom is also a member of the Pea family.*

More ... Broom was one of several species used to make 'brushes' to sweep and clean floors long before commercial brushes were produced.

Brambles, Briars and other wild Roses such as Dog-rose become more obvious as they come into flower. Typically scrambling through the woody shrubs of the hedge, they have a very significant presence in many hedgerows.

Although there is a **Dog-rose** 'species', many hybrids are formed with other closely related Rose species. Flowering is usually from May to July and pollination generally involves a wide range of insect species. Flowers can be pink or white or anything in between. The spines on the typically arching stems are normally curved backwards.

Dog-rose photographed in mid-June. Note the five petals and numerous stamens. Leaves have two or three pairs of leaflets, each with a toothed edge as seen in the photograph. These shrubs produce rosehips in autumn.

Moving on to the herbaceous plants of the banks and verges, there are many species which become much more obvious by May and June.

Creeping Buttercup, by far the most common of the Buttercup species in hedgerows, spreads rapidly filling gaps due to its system of runners which spread across available ground rooting at nodes to form new plants. It is much more robust than Meadow or Bulbous Buttercup, having denser foliage compared to the more delicate and deeply dissected leaves of the other two species. Flowering from May to September, Creeping Buttercup is common in hedgerows (particularly in disturbed sections), at the edges of paths, in gaps in woodland, on waste ground and even in gardens. It is much more of a 'generalist' in terms of habitat preference than other Buttercups.

A dense stand of Creeping Buttercup, photographed in early June. The flower stalks, each with a single terminal flower, extend above the dense leaf layer to a height of up to 50 cm.

Cow Parsley is the classic herbaceous hedgerow species and is at its peak in May (although also in flower for parts of April and June), typically covering large sections of hedgerow banks and road verges in a mass of lacy white blossom, reaching heights of up to one metre. Cow Parsley is perennial and overwinters as a taproot. Its finely divided and pinnate (fern-like) leaves are easily recognisable.

More ... Cow Parsley is a tall, fast-growing species which outcompetes and displaces shorter, slower growing species. One reason for this is that it can take advantage of the nutrients provided by hedge and grass clippings cut and left to decompose.

Cow Parsley dominating both hedgerow banks of this south Antrim minor road in mid-May.

The individual flowers are five-petalled, and as with Hemlock Water-dropwort, there are two levels of umbel structure producing both primary and secondary umbels.

This photograph (taken in late May) shows twelve terminal umbels of five-petalled white flowers in this Cow Parsley plant – individual flower stalks and the origins of the terminal secondary umbels are hidden behind the flowers in this photograph.

Cow Parsley is a member of the Carrot family and is a rich source of pollen and nectar for bees, butterflies and other insects.

May is also a time when **Greater Stitchwort** is at its best, although often in flower during April and June too. Greater Stitchwort typically scrambles through other vegetation in the hedgerow and produces clusters of flowers, each with five deeply divided white petals and ten stamens topped by yellow anthers. The white petals are of a brilliant white shade which is especially striking against the shade of the taller plants through which it climbs.

Greater Stitchwort photographed at the end of April. Common across most of NI, note the deeply divided petals, giving the impression that there are ten rather than five petals. There is a close-up of a flower of this species in the chapter on flower structure.

More ... It used to be thought that Greater Stitchwort was a cure for 'stitches' in the side.

While large colonies of **Early-purple Orchids** can be found in some deciduous woods, small groups or individual plants appear in some hedgerow banks and verges. As with Greater Stitchwort, the Early-purple Orchid can be in flower from late April to early June but is typically at its best in early May.

The Early-purple Orchid with its heavily spotted leaves organised as a rosette around the base of the stem and an inflorescence of flowers which are almost uniformly purple in colour. A more detailed photograph of the inflorescence of this plant is in an earlier chapter.

More ... The leaves of the Early-purple Orchid in the photograph are heavily spotted. This plant was photographed on a hedgerow bank. However, leaf spotting can be variable in this species and plants growing in grassland habitats can have unspotted leaves. Note the unspotted leaves of the Early-purple Orchids on the cover of this book.

More ... The Early-purple Orchid does not grow in dense shade and is normally found with low growing species such as Primrose, Violets and Wild Strawberries. These are also indicators of ancient woodland.

By June, the well-known annual, **Cleavers** (**Goosegrass**), is in flower and can be seen scrambling through shrubs and herbs in the hedgerow. Narrow leaves are in whorls of 6-8 at intervals along the stem. Flowering from May through to August, the stems and leaves have numerous minute prickles which enable the plants to work their way up and through the supporting shrubs and taller herbaceous plants.

Note the leaves in whorls at intervals along the stem and that branches originate at the same point as leaves. As can be seen, the flowers are in branched clusters. Very similar in structure to the closely related Woodruff (covered in the chapter on woodlands), both belong to the Bedstraw family.

Ground-ivy, Herb-Robert, Wild Strawberry and Barren Strawberry continue in flower for much of this time. Common Nettle comes into flower in mid-Summer. Woodland species such as Primrose, Wild Garlic, Wood Anemone and Bluebell are also common hedgerow species during these months although several of these have passed their peak flowering times by mid-Summer.

July, August, September and beyond

By July the banks and verges of many hedgerows have been cut, a consequence being the loss of the leaves and flowers of the plants of many species. Species with leaves in ground-hugging rosettes, such as Dandelions, can escape the blade as can those protected through growing close to the base of the barrier species. Additionally, in those verges not subjected to cutting, Grass species typically become dominant by mid-Summer and therefore may swamp many of the rarer and more slow growing species.

From June on, the more robust **Hogweed** replaces Cow Parsley as the dominant herbaceous species in many hedgerows. A native species, and much smaller than its close relative, Giant Hogweed, Hogweed seldom grows more than two metres in height and the terminal umbels are usually less than 20 cm in diameter in total.

Flowers are white to pinkish white, with the petals noticeably indented on their outer edge.

More ... Before toys were as readily available as today, and children were more likely to improvise, the hollow stems of Hogweed were used as pea shooters.

Hogweed photographed in early July. Note the large, wavy leaves, very different to the feathery leaves of Cow Parsley. Although difficult to see in this photograph, the outer petals in each umbel are indented.

A climber whose height in hedgerows is only limited by the height of its supporting shrubs, **Honeysuckle** (also known as Woodbine) is in flower from June through to September. The easily recognised flowers have exceptionally long (3-4 cm or longer) trumpet-like corolla tubes at the base of which is nectar. The very long corolla tube ensures that only those species such as butterflies, moths and bees, with the appropriate feeding tubes can reach the nectar deep in the base of the flower. The five stamens and single style extend well beyond the outer edge of the corolla, and the flowers are arranged in circular heads, as shown in the next photograph.

More ... Unusually for insect pollinated flowers, Honeysuckle produces more scent in the evening and night than during the day. This is to attract night-flying moths which are important pollinators.

Flowerhead of Honeysuckle. Note the very long corolla tube and the upturning of the upper lip and downward curling of the lower lip of each flower. The stamens and style (with green stigma at the tip) of each opened flower extend well beyond the corolla.

More ... The stems of Honeysuckle wind around the woody branches of the barrier species, often restricting their growth.

Tufted Vetch is one of the commonest hedgerow species in mid-Summer, typically scrambling through other herbs and even shrubs to gain height and light. Its leaves are very obviously pinnate, with between 6-15 pairs of narrow leaflets arranged directly opposite to each other on a common stem. Tufted Vetch produces 10-25 blue to violet flowers from June to September, originating mainly from one side of the spike. It is a member of the Pea family.

Tufted Vetch photographed in early August. Note the pinnate leaves. At the end of the leaves are tendrils which curl around other plants enabling the plant to climb through the vegetation.

More ... The individual flowers of Tufted Vetch look like a line of hummingbirds feeding on the stem.

From mid-Summer to Autumn, **Yarrow** is a very common species in hedgerow banks and road verges, particularly if the grass is not too long. Growing to a height of 50 cm or less, the clusters of dense, flat flowerheads are umbel-like in appearance. The small flowerheads have creamy-yellow disc-florets and white or whitish-pink ray-florets.

With its composite flowers, Yarrow belongs to the Daisy family. It is also very common in dry grasslands where it is not outcompeted by vigorously growing Grasses.

Yarrow photographed in late September in the relatively short grass of a hedgerow bank. Note the umbel-like flat flowerheads. The delicate feathery leaves of this species are quite distinctive.

More ... The scientific name of Yarrow is *Achillea millefolium*. Millefolium can be translated as a thousand leaves.

Ivy often gets a bad press but flowering so late in the year it can be a life saver to a range of insects as a source of nectar. It typically flowers from September to November or even into December. An evergreen climber, Ivy can grow profusely covering the stems of many hedgerow trees and shrubs. The plant has yellow-green, five-petalled flowers, each with five stamens and they are arranged in umbels as shown in the next photograph.

Ivy photographed at the end of September. Note the spherical umbels produced by the supporting flower stalk of each flower radiating out from a common point at all angles rather than the more restricted (upward) spreading of rays in the umbels of the Daisy family.

Photographed in mid-Winter, this image shows how Ivy can be a prolific producer of berries at a time when relatively few other fruits are available for birds. Overall, an exceptionally valuable plant for wildlife, Ivy also provides autumn nectar and winter cover.

Ivy species have their own family, the Ivy or *Araliaceae* family. Ivy is not restricted to hedgerows or to its role as a climber, being frequently present as ground cover in both coniferous and deciduous woodland.

More ... Ivy uses 'suckers' to stick to the stems and branches of trees as it climbs. Other species use different strategies. Honeysuckle winds around the stems and branches of shrubs and trees (as does Bindweed). Tufted Vetch uses its tendrils to attach to other plants as it climbs up through other vegetation

More ... Ivy suckers simply hold on and don't extract moisture or nutrients from the tree as some people imagine when they feel the need to cut Ivy off at the base of trees. It can though be a risk to more mature trees that are weakening for other reasons. Bulky ivy growth in trees can catch winter winds and bring trees down that might otherwise have survived had they not been covered with ivy.

Late Summer and Autumn is the time when a wide range of fruits develop in hedgerows. The ovules within the ovaries develop into seeds if they become fertilised and the ovary walls themselves develop to form fruits. Seeds dispersed by explosive action have already been mentioned (for example, Himalayan Balsam), but many fruits (and the seeds within them) in hedgerows are dispersed by becoming colourful and fleshy so that they are attractive foods for birds.

Many hedgerow trees, shrubs and plants produce fruits dispersed in this way, including Blackthorn, whose fruit is known as sloe.

Sloes of Blackthorn, photographed in early August.

More ... Sloes are used to make sloe gin, a popular homemade drink. Blackthorn has traditionally been used to make walking sticks and Irish 'shillelaghs'.

The haws of Hawthorn and the berries of Honeysuckle are easily spotted by birds seeking a nutritious meal and the birds become unknowing distributers of seeds well away from the parent plants.

Hawthorn haws

Honeysuckle berries

Some species are dispersed by producing fruits with small prickles which catch and stick to the fur of animals or even people's clothing. Examples include Cleavers and Burdock.

Prickled fruits of Cleavers

Burdock

Hedgerows should not be considered as homogenous habitats. The closely cropped hedgerows found in much of east County Down and south Antrim are very different to the sprawling hedgerows, often with ditches, found in the west of NI. The presence of a water filled ditch can add significantly to the biodiversity of a hedgerow with species such as Wild Angelica and Meadowsweet being present.

Aspect is important in hedgerows, with south facing banks often having a different range of species than corresponding north facing banks. Additionally, the same species can come into flower several weeks earlier on south facing banks when compared to north facing banks at the same location.

Ireland is one of the least wooded countries in Europe, yet in parts of the countryside where fields are small and hedges well-endowed with trees it can still feel very well wooded. There is no doubt that hedgerows are a reservoir for many declining woodland species and they also operate as wildlife corridors linking small, relatively isolated copses and small woods dotted around the countryside. Certainly, in many parts of NI, hedgerows provide an interesting range of habitats worthy of study.

In Summary

It is worth reiterating that while this book has covered most major plant habitat types in Northern Ireland, it has not covered anything approaching a majority of the wildflower species present in any habitat. For the habitats covered, some of the most common species are included, as are a selection of the rarities.

Some major groups have barely been touched on, including members of the Grass family (*Poaceae*), the Sedge family (*Cyperaceae*) – the Sedge family has around one hundred species in Ireland alone – and the Rush family (*Juncaceae*). Each of these families has many species, but they are difficult to identify, and while technically flowering plants, they may not retain the interest of most non-specialists.

More … Grass stems have compartmentalised sections, with leaves arising at junctions between compartments, whereas Sedges have solid stems which are usually triangular in cross section. Rushes have stems which are round in cross section and are typically associated with marsh and other wetland habitats. In all three families, wind pollination is typical and the flowers are normally unshowy.

More … There are many versions of the following rhyme to help differentiate between the three groups: 'Sedges have edges; rushes are round; and grasses have knees, right down to the ground'.

Common Sedge in flower, photographed in mid-June, showing a female flower with a mass of cream-coloured stigmas arising from a terminal inflorescence. Probably the commonest Sedge in the British Isles, typically occurring in marshes and wet grasslands.

Classification of flowering plants

Species covered have been given their English name(s) for simplicity. As many species have historically garnered several names, for consistency the names used are the primary names used in *New Flora of the British Isles* (Fourth Edition) by Clive Stace, 2019, C&M Floristics, regarded by many as the authoritative account of wildflower nomenclature. There are a few exceptions to this. For example, Wild Garlic is named as Ramsons in Stace, but Wild Garlic is the name more widely known in NI, so it is used here. Additionally, allocation of species to families is as in Stace.

Many experienced botanists prefer to refer to plants by their scientific or Latin names for the very reason that there is a lack of consistency in their common and local names. Take for example **Cleavers**, a plant known to many as Goosegrass or Robin-run-the-hedge, which has many other vernacular names include Hayriff, Sticky-back, Sweethearts, Kisses, and Catchweed. However, there is one scientific name, the binomial (two-part) *Galium aparine* for Cleavers. No other plant species has this scientific name, which provides clarity. The second part of the binomial name '*aparine*' is what is called the species name, and the first part '*Galium*' is the genus name. There can be one or several species in a genus, depending on how closely related a species is to its closest relatives. Using the example of Cleavers, other very closely related species sharing the same genus include species we have met in this book such as Lady's Bedstraw, Woodruff, and Heath Bedstraw (*Galium verum*, *Galium odoratum*, and *Galium saxatile* respectively). In summary, a genus is a grouping of very closely related species.

Another species with several names is **Gorse**, commonly known as Whin in much of NI and Furze in much of the south of Ireland. How many have contemplated whether Gorse and Whin are the same or different plant species? At least the binomial *Ulex europaeus* leaves no doubt and having a scientific name is invaluable in an international context.

Genome sequencing and other molecular advances have resulted in a re-evaluation of the relationships between and within flowering plant groups. For example, **Lesser Celandine** has historically been in the same genus as the Buttercup species we have come across earlier in this book – all within the *Ranunculus* genus and part of the *Ranunculaceae* (Buttercup) family. This can be seen by their scientific names of *Ranunculus ficaria* (Lesser Celandine), *Ranunculus acris* (Meadow Buttercup), *Ranunculus repens* (Creeping Buttercup) and *Ranunculus bulbosus* (Bulbous Buttercup). In recent years, Lesser Celandine has been moved out of the *Ranunculus* genus and has been renamed as *Ficaria verna*. While this amendment has been based on molecular data, observation of the species does show that Lesser Celandine is an outlier in some respects – the three Buttercups each have five sepals and five petals in their flowers, whereas Lesser Celandine has three sepals and an indeterminate number of petals. With these physical differences one could ask why they were ever in the same genus in the first place!

Readers will have picked up that the number of plant species in a family can vary enormously. We have come across several families with only one species. For example, Flowering-rush is the only member of the *Butomaceae* (Flowering-rush) family, while others such as the Daisy family contain vast number of species. Why should this be the case? Families with large numbers of species arise because they have many species with only small differences between them and they all fit the overarching criteria for that family. A family of one species arises because that species is unique in the sense it has features which are significantly different to all other species (and families) of wildflowers.

While this book is not intended as an identification guide per se, it may help non-experts identify the species included. Hopefully, it will help build an understanding of some of the most common families and the common features within some of those families. For example, a common feature among members of the Carrot family is the presence of umbels. Another example is that each plant in the very large Daisy family has a composite flowerhead with numerous small flowers or florets packed tightly together in a relatively flat flowerhead held together by bracts organised in an involucre. Many will easily recognise the 'pea-like' flowers associated with members of the Pea family and appreciate that a species with this type of flower is very likely to be a member of this family.

Obviously, some species have that 'Wow' factor when you come across them for the first time, or when you find an unblemished version in pristine condition. The Bee Orchid is an obvious example. However, there are very few wildflower species that do not have something new or interesting to offer as hopefully shown in the next section.

The Common Nettle – more to it than you think!

Think of the humble **Common Nettle** (better known to some as the Stinging Nettle). 'Common' as this species is the most common Nettle species in NI; the Small Nettle also occurs but is much rarer. The Common Nettle tends to occur most frequently in areas which have been used by humans, and where the ground is fertile (with high phosphate and nitrate levels), often through enrichment by the manure of domestic animals, or where phosphate-rich ash has accumulated from fire. A species which can grow to well over one metre in height, the Common Nettle typically grows in dense clumps in hedgerow verges, woodland, and waste ground as a result of rhizomatous spread and seed production.

Toothed leaves are arranged as opposite pairs with successive pairs along the stem being at right angles to the pair below and above, as is typical for members of the Dead-nettle family.

Plants are of separate sexes with the small, easily missed, flowers being organised in catkin-like spikes. Flowering is from June to September with the flowers being wind pollinated.

The Common Nettle – note the leaves in opposite pairs with alternative pairs being at right angles to the next pair. Single sexed flowers are in catkin-like branched spikes which originate from leaf axils.

The next photograph, taken from directly above a Nettle plant shows how the leaves orientate themselves in a way which maximises the light that reaches the plant leaves. To do this they extend the length of the petioles (leaf stalks) of lower leaves, moving the main part of the leaf further away from the stem. This of course is something that most plants will do, the outcome being that there is as little self-shading as possible.

Common Nettle showing how the leaf arrangement maximises the area of leaf surface receiving direct light.

Nettles have an important role in nature's **food webs**, being a host plant for caterpillars of a wealth of butterfly species and of course the insects and birds which feed on them.

Nettles, like many other species, have a long history of use in **traditional medicine**. Nettles have a range of antimicrobial chemicals – no doubt having evolved over a long time to protect them against bacterial and fungal attack –and this fact is important in their use in herbal medicines. Traditional medicines involving Nettle have been used in treatment for a range of conditions including arthritis.

As a **food**, Nettles have been used as a primary ingredient in soups and herbal drinks. Many species in the Dead-nettle family such as Lavender, Mint and Thyme are widely used in the kitchen.

In eastern Europe and other areas, Nettles have been used in the manufacture of **clothing** and other textiles. A very fibrous plant, it has the advantage of growing in abundance in the cooler climates associated with much of northern continental Europe. The stems are particularly fibrous, and clothes through the generations have been made from a range of blended products including Nettle and other materials such as wool and cotton.

When we think of Nettle, for many of us the first thing that comes into mind is getting a Nettle **sting** as a child. We are all familiar with the toxin produced by the Common Nettle should we brush against its leaves. Upon contact, the tips of tiny hairs called trichomes snap leaving a needle-shaped structure which can penetrate the skin causing the release of histamine, acetylcholine and other chemicals which irritate skin causing it to become inflamed. The trichomes have probably evolved as an important defence against grazing animals.

For generations and to the present day, children who have been stung by Nettles rubbed the large leaves of Broad-leaved Dock on affected parts, often reciting the following rhyme:

'Docken in, docken out
take the sting of the nettle out'

Surprisingly, any relief is thought not to be due to any chemical found within Dock but is thought to be a combination of the cooling affect brought about by its sap and perhaps more importantly its placebo effect.

More ... Broad-leaved Dock traditionally has had other roles in addition to acting as an antidote to Nettle sting. Its broad and thick leaves were used to wrap blocks of butter to keep the butter cool in the pre-fridge era. For this reason, in many parts of the country it was known as 'Butter Dock'.

The Common Nettle is of course not unique in being a common, yet often ignored, plant with an interesting history. This little detour into the natural history of the Nettle is to emphasise this very point – most wildflower species do have a story to tell!

Fruits and Seeds

There has been little focus on what takes place after the processes of pollination and fertilisation, except for a short section in the previous chapter on hedgerows. This is not to say that a study of seeds and fruits cannot be very interesting and well worth further study, but it is not a focus of this book. Nonetheless, it is worth pointing out that in many species the fruiting stage can be even more interesting than the flowering stage. Wild Carrot, a member of the Carrot family, with the classic umbel arrangement of that family, and a species common on coastal grasslands is a case in point. Not a particularly distinctive species when in flower (from June to September), the concave 'birds-nest' appearance of the primary umbel when in fruit is perhaps more interesting and eye catching as seen in the next photograph.

Occasional reference has been made to the role of a particular species in herbal or traditional medicine. Many more such links could have been made as many species have a history of use in this role. Not particularly surprising, as in the centuries before modern drug development and pharmacology, herbal medicine was one of very few options available, irrespective of how tenuous the link between herbal concoction and recovery.

Global warming is something which affects us all. Reference has been made to how increasing temperatures appear to be affecting the ranges of plant distribution. The Bee Orchid was used as an example in an earlier chapter, and this is only one of many examples which could have been used. Furthermore, there is increasing evidence that the flowering periods of many species are also changing in line with the changing climate. Many Spring species are flowering earlier, often much earlier. A worry is that a change in the life cycle timeline of a plant species may not match a change in the timeline of its pollinators.

Most of the photographs used in this book were taken in 2021, a strange year when many of us led very different lives to what we deem to be normal. It was a year which presented the opportunity for so many to get outdoors and take a fresh look at nature. Many of the photographs used in this book have an interesting history. For example, the photograph of the Green-winged Orchid in the Introduction was taken at around 7.30 in the morning. This timing was to have the rays of the rising Sun at a low angle to highlight the delicate, green-veined sepals of this species. As many photographers say, light is everything!

OPPOSITE: *Inwardly folded secondary umbels of Wild Carrot during the fruiting stage, photographed in late August. Individual fruits are covered by short bristles and are at the end of relatively long secondary rays, clearly seen to be uniting at the same points of origin on their supporting (primary) ray.*

Anyone interested in flowering plants, whether in the garden or in the wild will have a favourite or group of favourites. It's hard not to. Orchids are top of my list, and I am clearly not unique in this. It's not difficult to work out why – many species are very difficult to find in the wild, and the susceptibility of many Orchid species to habitat change highlights their vulnerability. Their 'wow' factor as highlighted by this final photograph of the Common Spotted-orchid really cannot be overstated.

The Common Spotted-orchid photographed in mid-June in a species-rich grassland.

Glossary

Acidic (soils) – soils with a low pH. Acidic soils have low levels of basic (essential) minerals. Soils on rocks such as granite or sandstone tend to be acidic as are peaty soils.

Alkaline (soils) – soils with a high pH (as opposed to acidic soils which have a low pH). Alkaline soils normally have high levels of basic (essential) minerals. Calcareous soils and soils on chalk and limestone are typically alkaline.

Annual – a plant which completes its life cycle in one year.

Anther – the part of the stamen which produces pollen.

Axil – the upper angle between a leaf (or leaf stalk) and the stem from which it originates.

Basal leaves – leaves arising from the base of the stem at ground level.

Base-poor (soils) – soils which are low in alkaline minerals such as calcium and magnesium.

Base-rich (soils) – soils which are rich in alkaline minerals such as calcium and magnesium.

Biennial – a plant which completes its life cycle in two years. Biennials usually only flower in their second year.

Bilateral symmetry – there is only one plane of symmetry. Bilaterally symmetrical flowers can only be sliced into two equal halves in one vertical plane, for example, Gorse and Orchids.

Binomial name – the scientific or Latin name given to a species. Each plant species has one binomial name only which consists of a genus and species name.

Bract – a modified, scale-like 'leaf'. Often protective in function, but may have a role in attracting pollinators.

Bulb – an underground swelling at the base of a stem. Plants may produce a main bulb but also daughter bulbs. Daughter bulbs may detach from the parent plant to produce new plants by vegetative propagation (asexual reproduction).

Calcareous (soils) – soils which are rich in calcium carbonate (lime).

Calyx – the collective term for the (whorl of) sepals in a flower.

Carpel – the female part of a flower. Includes the ovary, style and stigma. Usually, the central component of a flower.

Catkin – a spike of small reduced (or unshowy) flowers, typically of the one sex.

Composite (flowerhead) – a flowerhead of a member of the Daisy family. Also describes a plant belonging to this family.

Corolla – the collective term for the (whorl of) petals in a flower.

Cross-pollination – the pollination of a flower by pollen transferred from the flower of another plant (of the same species).

Disc-floret – a tubular floret in a composite flowerhead.

Family – a grouping of related species which possess many similarities. All members of the same family have more features in common with each other than they do with species belonging to other families.

Fertilisation – the fusion of a male and female gamete.

Filament – the stalk of a stamen which supports the anther.

Fruit – the ovary post-fertilisation, which contains one or more seeds.

Gamete – sex cell. In flowering plants gametes are found within pollen grains and ovules.

Genus – a grouping of closely related species. A number of genera are typically grouped into a family.

Herbaceous – non-woody (green) plants. The above ground parts of herbaceous plants typically break down and decompose in winter.

Heterostyly – situation where there are two distinct forms of flower for the same species. Normally seen in the relative positioning of anthers and stigma.

Hybrid – a plant arising from cross breeding between members of two different species. Normally, hybrids only form between very closely related species.

Inflorescence – a grouping of flowers in a particular structural format. More organised than a **cluster**, where a localised grouping of flowers is not organised in any particular pattern.

Involucre – a whorl of bracts at the base of a flowerhead (for example, in composite flowers).

Ovary – the part of the carpel which contains the ovule(s).

Ovule – the 'egg' or structure which contains the female gamete. Found within the ovary of carpels.

Panicle – a branched raceme.

Parasite – a plant which gains its nutrients from another plant (and causes the host plant a degree of harm – such as by 'stealing' some of its nutrients). Parasites can be 'total' as with Toothwort, or 'partial' as with Yellow-rattle. Hemiparasite = partial parasite.

Pedicel – the stalk of an individual flower.

Perennating organs – structures (usually underground such as bulbs and rhizomes) which allow a plant to overwinter and survive from one growing season to the next.

Perennial – living for more than two years.

Perianth – the collective term for the calyx and corolla combined (includes all sepals and petals). A term that tends to be used when it is difficult to distinguish between sepals and petals.

Petiole – the stalk of a leaf.

pH – a measurement of the acidity or alkalinity of soil or water.

Pinnate leaf – a compound leaf with leaflets arising along a leafstalk, usually in opposite pairs and typically with a terminal leaflet.

Pollen – minute grains which are transported from anther to stigma in pollination. A pollen grain contains a male gamete.

Pollination – the transfer of pollen from anther to stigma.

Raceme – an elongated inflorescence in which the flowers open in sequence from the bottom of the inflorescence to the top.

Radial symmetry – a flower in which the component parts are arranged regularly around a central point. A radially symmetrical flower does not have a left and right side.

Ray – a branch in an umbel. Typically, used to describe the branches leading from the common stem to secondary umbels.

Ray-floret – strap-shaped floret (usually in an outer ring) in a composite flowerhead.

Rhizome – an underground (usually horizontal) stem. Often a food storage structure but important in lateral vegetative spread.

Rosette – a flattened whorl of leaves organised as a ring around the base of a plant's stem.

Runner – an above ground horizontal stem which creeps across the ground, rooting at intervals to form new plants.

Saprophyte – a plant which does not photosynthesise, but obtains its nutrients by absorbing decomposed organic material.

Seed – a fertilised ovule.

Self-pollination – pollination in which pollen is transferred between anther and stigma in the same flower, (or between the anther of one flower and the stigma of a different flower in the same plant).

Sepal – part of the outer whorl in a flower. The ring of structures immediately outside the petals. Often green and normally protective in function, but may be coloured to help attract insects.

Shrub – a much branched woody plant, without a main stem (as in a tree). Shrubs do not reach the size of trees at maturity.

Species – a type of plant (or organism). All members of a species closely resemble each other and are normally capable of interbreeding with each other to produce viable offspring.

Spur – a tubular projection from a petal or a sepal, within which nectar is made and contained.

Stamen – the male part of a flower. Includes the filament and anther.

Stem leaves – leaves arising from the stem (as opposed to basal leaves). In many plant species, the stem and basal leaves are distinctly different.

Stigma – the (distal) part of a carpel which is involved in pollination.

Style – the section of the carpel between the ovary and the stigma. Its length is often important in ensuring that the stigma is in a suitable position for pollination and/or encouraging cross-pollination.

Sub-species – sub-species (of the same species) can breed together. However, there are minor but significant differences between the members of one sub-species and another sub-species (of the same species). Only some species have sub-species.

Succulent – fleshy due to large volumes of stored water.

Taproot – the main root in a root system, which normally grows directly down from, and is continuous with, the main stem. Usually considerably thicker than the other (lateral) roots. Lateral roots branch from the (usually centrally positioned) taproot.

Tepal – one of the segments in the perianth. The term tepal tends to be used when the sepals and petals are similar in appearance or if the flower contains only one of the sepals or the petals.

Trifoliate leaf – a compound leaf with three leaflets.

Tuber – a swollen, fleshy underground structure of stem or root origin which acts as a food storage unit. Tubers may detach from the parent plant to produce new plants by vegetative propagation (asexual reproduction).

Umbel – a (usually flat-topped) inflorescence in which all the pedicels (flower stalks) arise from the same place. Compound umbels are formed when there are secondary umbels in addition to a large primary umbel on each flowering stalk.

Index of Plant Species

Page numbers in **bold** indicate a photograph of the plant.